Corporate Management
of Health and Safety Hazards

Corporate Management of Health and Safety Hazards

A Comparison of Current Practice

Roger E. Kasperson,
Jeanne X. Kasperson,
Christoph Hohenemser,
and Robert W. Kates
with Ola Svenson

Westview Press
BOULDER AND LONDON

Westview Special Studies in Science, Technology, and Society

This Westview softcover edition is printed on acid-free paper and bound in softcovers that carry the highest rating of the National Association of State Textbook Administrators, in consultation with the Association of American Publishers and the Book Manufacturers' Institute.

Published in 1988 in the United States of America by Westview Press, Inc., 5500 Central Avenue, Boulder, Colorado 80301, and in the United Kingdom by Westview Press, Inc., 13 Brunswick Centre, London WC1N 1AF, England

Library of Congress Catalog Card Number: 88-10619
Cataloging-in-Publication Data available
ISBN 0-8133-7615-7

Printed and bound in the United States of America

The paper used in this publication meets the requirements of the American National Standard for Permanence of Paper for Printed Library Materials Z39.48-1984.

10   9   8   7   6   5   4   3   2   1

Contents

List of Tables   *vii*
List of Figures   *ix*
Preface and Acknowledgments   *xi*

1   Introduction   1

The Study Sample: Large and Wealthy Corporations   4
An Industry-Wide Perspective   7
Conceptual Approach   9
Organization   13

**2   Managing Hazards at PETROCHEM
       Corporation**   15

The Structure of Hazard Management   16
Product-Safety Management   26
Occupational Hazard Management   32
Conclusions   39

**3   Managing Occupational Hazards at a
       PHARMACHEM Corporation Plant**   43

Production of PRODUCT-A   43
Hazards of PRODUCT-A Production   44
Control Actions for the Hazards of PRODUCT-A   44
Occupational Hazard Management   45
Conclusions   54

4   **Managing Product Hazards at Volvo**
        Car Corporation, *Ola Svenson*   57

        Hazard-Management Organization   58
        Safety of Volvo Cars and Company Behavior   62
        Hazard Management and Quality Control at Volvo   69
        The Safety-Information Feedback System   70
        Conclusions   76

5   **Managing Occupational and Catastrophic**
        Hazards at the Rocky Flats Nuclear
        Weapons Plant   79

        A History of Controversy   80
        A Tour of the Plant   82
        Occupational Health at Rocky Flats   87
        Assessing and Managing Catastrophic Hazards   93
        Summary and Conclusions   97

6   **Union Carbide Corporation**
        and the Bhopal Disaster   101

        Union Carbide Corporation and Its Hazard-Management
            Programs   104
        Union Carbide Comes to Bhopal   105
        Early Warnings   107
        The Accident   108
        Corporate Hazard Management   111
        Concluding Note   117

7   **Corporate Management of Health and**
        Safety Hazards: Current Practice and
        Needed Research   119

        Characteristics of Current Practice   119
        An Agenda of Needed Research   126
        The Future of Corporate Hazard Management   131

        References   133
        Index   141

Tables

1.1	Overview of Case Studies	5
1.2	Vital Statistics of the Corporations/Plants Studied	6
1.3	A Seven-Class Taxonomy	13

| 2.1 | OSHA Permissible Exposure Limits (PELs) and PETROCHEM Internal Standards | 22 |
| 2.2 | PETROCHEM Corporation | 29 |

| 3.1 | Potential Chemical Hazard Releases in the Production of PRODUCT-A | 46 |
| 3.2 | Control Actions by Hazard Release and Causal Stage | 48 |

| 4.1 | Some U.S. and Swedish Legal Requirements Compared with Volvo Standards as of 1981 | 65 |

5.1	Major Accidental Releases of Plutonium from the Rocky Flats Plant	81
5.2	Some Plutonium Hazards and How They are Controlled at Rocky Flats	88
5.3	Internal Body Burden of Plutonium at Rocky Flats	92
5.4	Summary of Postulated Higher Risk Accident Scenarios	95

| 6.1 | Significant Industrial Disasters of the Twentieth Century: A Chronology | 103 |
| 6.2 | Workplace Limits to Chemical Exposures | 106 |

Figures

1.1   The hazard chain model   10
1.2   A flow chart of hazard management   12

2.1   PETROCHEM Corporation: Statement of Philosophy   18
2.2   PETROCHEM's risk-region approach   24
2.3   Matrix management at PETROCHEM Corporation   27
2.4   PETROCHEM Corporation: Health, Safety, and Environment   28
2.5   The PETROCHEM computerized health-surveillance system   34
2.6   Occupational injury/illness experience manufacturing   37
2.7   Total recordable incidence rates   38
2.8   Incidence rates - Fatalities plus cases with days away from work   40

4.1   Organizational structure of Volvo   60
4.2   Investigative matrix for automobile and accident research   66
4.3   Volvo's feedback system   67
4.4   Feedbacks enabling improved safety during planning, production,
         marketing, and use of a car   72

5.1   Plutonium-239 contours around Rocky Flats   86
5.2   A diagram of the causal structure of hazard   89
5.3   Whole-body external radiation exposure among Rocky Flats
         workers since 1953.   90
5.4   Comparison of rare event risks   96

6.1   Extent of MIC dispersal at Union Carbide's Bhopal site   102
6.2   MIC tanks at the Bhopal pesticide plant   109
6.3   The causal structure of hazard: Application to the Bhopal
         accident   110
6.4   Detailed model of contributors to the Bhopal accident   112

Preface and Acknowledgments

Corporations have been the subject of much discussion during the 1980s. Mostly this attention has focused on the economic competitiveness of American industry or the degree of technological innovation achieved. The role of the corporation in managing the myriad technological hazards confronting society has received much less scrutiny. Indeed, as the pages that follow reveal, the corporate management of health and safety hazards is terra incognita.

This book aims at a better understanding of the current practice of hazard management by large corporations, particularly those that have emerged as industrial leaders. The focus is on how corporations go about the job of identifying and managing hazards, on the resources committed to the tasks and on the characteristics likely to breed or contribute to success or failure. Books that treat industrial success as well as failure are not popular. Since this study inquires into the actual practice of hazard management in large and economically successful corporations, we expect that many readers will miss tales of exposé and industrial abuses.

The heart of the volume contains five case studies of individual corporations and their hazard-management programs. Largely, the book comprises a collaborative effort by four researchers in the Hazard Assessment Group at Clark University's Center for Technology, Environment, and Development (CENTED), although certain individuals took the lead for particular chapters. Chapters 1 and 7 result from collaborative discussions among all four members of the group. Roger E. Kasperson was the principal author of chapter 2; Robert W. Kates of chapter 5, and Christoph Hohenemser of chapter 5. Jeanne X. and Roger E. Kasperson shared the writing of chapter 6 as well as the overall editing of the volume. For the analysis of the Volvo Car Corporation (chapter 4), we are fortunate in enlisting the effort of our Swedish colleague, Ola Svenson, who brings expertise both in risk analysis and decision theory.

We acknowledge a particular indebtedness to the Russell Sage Foundation for supporting this effort when other potential funding sources were reluctant. We appreciate immensely the willingness of the several corporations described herein to open their doors and documents for our inspection when most other firms refused or were hesitant. Unfortunately, the corporate officials who gave so generously, and patiently, of their time and expertise must remain nameless. To a number of Clark University staff members, particularly Patricia Auger, Joan McGrath, Betty Daughney, and Terry Reynolds, we are extremely grateful for the patience, good humor, and hard work with which they produced draft upon draft of the manuscript.

Above all, we salute Lu Ann Pacenka for seeing us through yet another volume. This time around she has taken a hand to composition and design. She selected Microsoft Word Software and a combination of Times and Helvetica fonts to produce the book with a Macintosh SE and a Laser Writer Plus printer. This handsome volume bears the indelible stamp of her own rare blend of enthusiasm, perfectionism, and style.

Roger E. Kasperson
Jeanne X. Kasperson
Christoph Hohenemser
Robert W. Kates

1

Introduction

The industrial-scientific revolution in the design, production, and use of technology had traversed two centuries before industrial societies turned to the comprehensive management of the technological hazards attendant on that revolution. Whether one dates the beginnings of this effort to the early warning popularized in Rachel Carson's *Silent Spring* in 1962, the mass public outcry of Earth Day in 1970, or the classic paper of Chauncey Starr (1969), the movement is less than a quarter century old. Discernible innovations in the way society handles technological hazards are under fifteen years old, but the profound nature of the changes in attitudes, institutions, and activities may well hail, in retrospect, a revolution no less impressive than its predecessor.

It is always difficult to take stock in the midst of rapid change. Studies have concentrated on two features—the emergence of the environmental, consumer, and personal-health movements and the societal response in the form of legislation, governmental agencies, and regulations that cover the varied domains of technological hazards. Recently, attention has focused on changes in judicial processes, their roles in monitoring administrative performance, and more important, in providing compensation to victims as well as in deterring negligence (Huber 1986). But perhaps the most profound and least-recognized changes have occurred in the corporate world.

That world's management of hazards is uncharted terrain. This is so partly because the corporate world has a deep-seated reluctance to submit to external inspection or to encourage introspection. Moreover, several corporate biases come into play. In graduate schools of business and in economics departments, students of corporate activity focus primarily on managerial strategies related to the "bottom-line" measures of profit and efficiency. Graduate schools of business harbor few professors of hazard management. Conversely, social scientists feel more comfortable with the unfettered research possibilities in studying the

1

more accessible domains of governmental or public behavior. Others see hazard management as the incidental outcome of conflicting forces in capitalist society. Finally, accounts of hazard management have dwelled upon corporate failures and abuses (Johnson 1985).

A prevalent view has it that hazard management in corporations is fundamentally flawed by the conflict between safety and short-term profit, that without the pressure of regulation, continuous public scrutiny, and liability law, corporations inevitably remain "bad actors" who will undermine public and occupational health and safety. Evidence of callous behavior is easy to document, particularly among smaller, "marginal" firms with minimal investments in, and concerns with, health and safety. Failures in hazard management show up in poorly educated, illegal immigrants who handle highly toxic chemicals in poorly ventilated work areas, workers exposed for years to asbestos or cotton dust, or "midnight dumpers" of chemical wastes that eventually contaminate ground water (OTA 1983). In other cases, companies have acted within the safety context that operated at the time, only to be greeted by the changed values and recriminations of a new generation of publics and workers.

But abuses and failures do not capture the whole spectrum of corporate hazard management. Some corporations not only have at their disposal extensive resources for hazard analysis and management but have conscientiously organized to do an effective job. For these corporations, government regulation is a continuing prod (and sometimes nuisance) that the internal apparatus for risk management must endure. Oftentimes, these corporations have risk-reduction and safety goals that exceed those embodied in government standards and that find support within the organization not only because they contribute to a safer workplace and healthier consumer but because they provide long-run stability, fewer liability threats, a good corporate image, and even (occasionally) a competitive edge over firms that lack the resources to mount similar efforts.

In short, corporate hazard management is a complex subject prone to over-simplification. Simplistic theories and stereotypes are likely to have limited application, and governmental regulation has different impacts and roles in different corporations. Thus, a timely examination of hazard management in large corporations is needed to profile how they actually go about the task of managing technological hazards and making trade-offs among conflicting objectives and constraints. Such empirical scrutiny can help to test the accuracy of prevailing assumptions and stereotypes, as well as to complement the already flourishing literature of comparative case studies of the public sector.

Then, too, structural considerations prompt an examination of corporate hazard management. Previous studies of 93 technological hazards (Hohenemser, Kates, and Slovic 1983) have suggested that the management of hazards usually works best at stages of design and choice of technology. Decision making at these points almost invariably lies in the hands of industry. Thus hazard makers are also potentially the most effective hazard managers. The rub comes in creating the reality of corporate responsibility in a milieu of corporate pressures to avoid hazard management costs and to externalize costs on society as a whole.

Finally, the scope of corporate investment—notably over the past decade—in hazard assessment and safety analysis holds particular promise. In many cases, these resources are so extensive as to parallel or exceed the resources available to government regulators. Enhanced public access to such resources would enlarge substantially society's total capability for dealing with hazards of all kinds. The chemical industry's massive investments in screening and testing chemicals, for example, would make an extensive addition to the public resources in the field. It is worthwhile, therefore, to explore potential pathways for gaining access to these corporate resources and for bringing industry research and findings, suitably sifted and separated from proprietary information, into the public domain.

Accordingly, five years ago the Hazard Assessment Group at Clark University's Center for Technology, Environment, and Development (CENTED) set out to explore the possibility of studying corporate hazard management. The group wet its feet with a site visit to a major chemical producer. But two years elapsed before the clearing away of other research commitments allowed the planning of a full-fledged study. The planning process soon ran up against the problems of securing funding and gaining access to corporations. Major public agencies were reluctant to disburse funds for the study of private industry, and the Clark group felt uneasy about tapping private industry for support to study private industry. Fortunately, a modest grant from the Russell Sage Foundation came through to permit initiation of the exploratory research.

The long-term goal aims to answer a pivotal question: Given the inherent conflict of interest in corporate roles between enlarging profit and securing safety and given the apparent limits to society's capability to regulate industrial hazards, how can society more effectively tap the competence and experience of industry in managing technological hazards? To begin to tackle this ambitious question, this exploratory study addresses three more modest objectives:

- to profile the current structure of hazard management in selected corporations
- to analyze in depth the generic institutional activities in which corporations engage and the extent of corporate resources that support such efforts
- to evaluate, where possible, the outcomes of these efforts, noting the sources of success and failure.

To foster a better understanding of the current practice of hazard management in large corporations, the following chapters present case studies of five specific corporations. The researchers sought cooperation from corporate officials and offered reasonable protection of the corporation's vital interests in areas of liability, proprietary information, and the like. In general, the research sought information on the organization, scope, priorities, and substance of hazard-management activity as well as the processes and resources for arriving at corporate decisions and resolving internal conflicts—buttressed, wherever possible, by empirical data and examples. The researchers, in return, strove to describe their findings objectively and in a comparative framework, to permit corporate review of (but not changes in) the analyses, and to refrain from judgments for which the data base was inadequate.

For three of the detailed case studies (MACHINECORP, PETROCHEM, and PHARMACHEM), the agreements stipulated that the corporation (or plant) and its location(s) remain anonymous. One of these companies (MACHINECORP) eventually decided to terminate its participation in the study. Two other cases (Volvo and Rocky Flats) carried no requirements for anonymity. One case (Bhopal) entailed extensive use of a voluminous public record as well as a visit to corporate headquarters and a subsequent meeting with corporate officials. The remaining cases involved site visits, briefings by hazard managers, or both.

A case study of the Chemical Manufacturers Association, field visits to Digital Equipment Corporation and Monsanto Chemical Company, and interviews with representatives of a regional insurance company rounded out the study. Our general discussion includes the results from these efforts: Table 1.1 provides an overview of the case studies and indicates the authors' collective judgment as to the level of understanding that the research achieved.

The Study Sample: Large and Wealthy Corporations

As Table 1.2 shows, the corporations studied are relatively large, with total employment ranging from 20,000 to 100,000, and annual incomes falling in the range of $.34-$1.6 billion. In three cases (Union Carbide, PHARMACHEM, and Rockwell International), the study focussed on individual plants with specific missions. Even these plants are large, with employment ranging from 400 to 3,700. All of the five corporations studied in detail are involved in modern technology in which innovation, automation, and (with the exception of the Rocky Flats) strong competition and profitability play dominant roles. None of the five corporations was in financial difficulty (although the Union Carbide plant at Bhopal was having problems), and one of the plants (Rocky Flats) operated on a government financed cost-plus basis with essentially no limit on hazard expenditures. Although the ages of the corporations, ranging from 40 to 99 years, indicate mature institutions, the younger ages of particular plants (12-34 years) reflect recent innovation. Even the oldest plant studied (Rocky Flats) has undergone extensive rebuilding and renovation in the past five years.

The study sample, then, is highly skewed toward one end of the spectrum of corporations. Clustered in this space are the leaders, corporations with state-of-the-art programs for managing hazards. Indeed, except for the Bhopal calamity, the sample is devoid of clear failures in management or "worst-case scenarios" come true. It is essential to bear this in mind.

Uneven access to corporations and to hazard managers rendered the five case studies somewhat incommensurate. For the three corporations (PETROCHEM, PHARMACHEM, and Volvo) that provided a high degree of access, the study yielded extensive information about the organization of hazard management and much less about the structure, consequences, and assessment of hazards. In the case of limited access (Rockwell International), the analysis focuses upon the hazard, its potential consequences, and risk assessment and presents relatively scant detail about the organization of hazard management within the corporation. For the Bhopal plant, it was possible to tap both a general understanding of the

Table 1.1

OVERVIEW OF CASE STUDIES

CORPORATION/ PLANT OR ASSOCIATION	LEVEL OF UNDERSTANDING OF CORPORATE MANAGEMENT				COMMENTS
	1 General survey of scope of hazard management	2 Schematic knowledge of organization, priorities, and activities	3 Substantial understanding of organization, priorities, and programs	4 In-depth under-standing of management system and outcomes	
Digital Equipment	X				Single plant visit
Monsanto Chemical	X				Briefings and discussions with corporate managers
Hanover Insurance	X	X			Briefings by hazard managers
Dow Chemical	X	X			Extensive briefings; visit to research facilities
MACHINECORP	X	X			Study terminated by company
Volvo	X	X	X		Cooperation throughout study
PHARMACHEM	X	X	X		Cooperation throughout study
PETROCHEM	X	X	X	X	Excellent cooperation and documentation on organization and accomplishments
Rockwell International/ Rocky Flats Plant	X	X		X	Access through governmental visiting committee
Union Carbide/Bhopal	X	X		X	Extensive data base on accident; field visit to corporate headquarters
Chemical Manufacturers Association	X	X	X		Cooperation throughout study; substantial documentation

Table 1.2

VITAL STATISTICS OF THE CORPORATIONS/PLANTS STUDIED

	Union Carbide/ Bhopal Plant	PETROCHEM	PHARMACHEM Plant	Rockwell International (Rocky Flats)	Volvo
PRINCIPAL PRODUCT					
Corporation	Chemicals	Chemicals	Pharmaceuticals	Weapons	Cars and Trucks
Plant	Pesticides	N/A	Four drugs	Plutonium weapon components	N/A
NUMBER OF EMPLOYEES					
Corporation	100,000	36,000	40,700	120,000	76,200
Plant	800	N/A	400	3,700	N/A
NET SALES					
Corporation	(UCC) $9.5B (UCIL) $175.0M	$20B	$3.8B	$11.3B	$13.0B
Plant	$14.0M	N/A	N/A	N/A	N/A
NET ANNUAL INCOME					
Corporation	(UCC) $340.0M (UCIL) $12.3M	$1.6B	$500M	$700M	$425M
Plant	<$0.0*	N/A	N/A	N/A	N/A
HEALTH & SAFETY WORKFORCE					
Corporation	750	250	N/A	N/A	>15
Plant	<20	N/A	<10	250+	N/A
AGE IN YEARS					
Corporation	(UCC) .99 (UCIL) 50	40+	N/A	50 or 60+	61
Plant	15	N/A	12	34	N/A

N/A = Not applicable or not available

* The Bhopal plant was losing money at the time of the accident.

corporate management structure and a detailed data base on the causes and management of an industrial disaster.

Perhaps the most striking example of successful industrial hazard management is the Volvo Car Corporation, which occupies a unique niche in innovating safety in the automobile industry. Long before Detroit awoke to the fact that safety sells, Volvo stood virtually alone in the industry in promoting and advertising a vigorous corporate effort in safety design. A pioneer in installing safety belts before they were required, Volvo introduced many other safety-design features before they became standard elsewhere in the industry. Volvo supports efforts to assure vehicle safety through a unique feedback system, an accident-investigation effort that sends company investigators scurrying to all accidents within a certain radius of company headquarters. The company had also integrated hazard management into overall corporate quality assurance. Organizationally, health and safety at Volvo play a role less visible and distinct than at PETROCHEM and Rockwell, but that is probably appropriate for an industry where product safety is best achieved through design and quality control.

Three corporations—PETROCHEM, Union Carbide, and Rockwell International—stand out as economically successful firms in chemical processing. PETROCHEM avowedly aspires to lead the petrochemical industry in the area of health and safety. Both PETROCHEM and Union Carbide sometimes set internal standards more stringent than federal regulations, support their efforts through extensive in-house health research, and affirm their overall organizational pursuits through formal implementation programs. Rockwell International—at least at the Rocky Flats Plant that it runs for the government—takes a comparable approach and, in addition, uses an extensive "defense-in-depth" approach to hazard management.

PHARMACHEM manages occupational safety by following well-established "standard procedures" and builds product health and safety into process management. Union Carbide, an acknowledged leader in industrial safety in the chemical industry, shows in the Bhopal accident how widespread failures in implementation, design, maintenance, and management can compromise sound principles of risk management and effect a colossal collapse of defense in depth. It is also apparent, in retrospect, that Bhopal reveals serious generic shortcomings in the management of toxic substances at Union Carbide and in the chemical (and other) industries more generally.

All five corporations tend to lie at one end of the spectrum of corporations in size and resources. Consequently, it is essential to view them against the broader and variegated industrial context.

An Industry-Wide Perspective

In 1982, the Chemical Manufacturers Association engaged an independent accounting firm to conduct a survey of the U.S. chemical industry to collect information about the efforts by individual companies "to reduce unreasonable risk to health and the environment" (Peat, Marwick, Mitchell and Co. 1983). Representing slightly more than half of the industry, as measured by U.S. sales and employment, 112 firms responded. The firms ranged in size from one with a single product, 37 employees, and $2 million in sales to one with 17,750 products, 67,000 employees, and $7.8 billion in sales. The composite firm (as described by the mean responses in the survey) manufactured 1,070 products, had 4,400 employees, and enjoyed $733 million in U.S. sales. For this composite firm, hazard management represented a substantial effort, although just how much is difficult to determine. It is nonetheless possible to extract from the survey data that characterize the chemical industry at that time, including employment, expenditures, and managerial attention to health and safety

The mean number of full-time-equivalent health and environmental specialists reported for 1981 was 84, representing about 2 percent of all employees. In turn, it is possible to estimate corresponding numbers of support staff by taking the overall figures for the entire staff for air, water, and solid waste pollution control and comparing these to the number of specialists (a multiplier of 3.45). Applied over all to the specialist estimate, this comparison yields an estimated mean of 290 full-time personnel, or about 6.5 percent of employees, involved in hazard management. Added to these are many specially trained part-time personnel—for example, corporate emergency-response teams for chemical spills averaged about 160 people.

The composite corporation incurred a 1981 bill of $1.7 million for toxicity testing and $25.6 million for pollution control (including capital investment), representing 3.7 percent of corporate sales. Using the estimated operating and maintenance cost per employee for pollution control of slightly more than $150,000, this would give an overall annual cost for the 290 estimated employees in 1981 of $43.5 million (or about 6 percent of annual sales).

Ninety percent or more of the firms reported formal internal policies dealing with product safety, worker health and safety, and environmental protection. Fully 84 percent had disaster plans of some sort (even prior to Bhopal). Yet only in 30 percent of the corporations responding were the director, the president, or both, routinely involved in health and environmental matters; in most other cases, a vice president handled these issues. This suggests that in only a minority of firms, despite all the recent management progress, were the highest levels of corporate decision making routinely involved in hazard issues.

The actual use of corporate resources may also be gleaned from the survey. Data on hazard-management functions suggest a relative corporate effort at the time of 11 percent for product safety, 23 percent for worker safety, and 31 percent for environmental protection, although the amorphous group of health, legal, and other specialties (35 percent) confounds the categorization. It is also apparent that all companies were keeping lists of all products and materials, and most companies were maintaining lists of pollutants as well. But few corporations claimed to review regularly the health records of their employees, again suggesting the widespread variability within the industry (particularly as compared with the case study of PETROCHEM).

A striking result from this industry survey is how recent these hazard-management activities are. The survey made no systematic effort to pinpoint the dates of all activities for which data were collected. Even so, it is clear that most hazard-management practices did not begin until 1972 or later, findings that tend to confirm the results of our case studies (with the noteworthy exceptions of Volvo and Rocky Flats).

Conceptual Approach

The case studies that take up the next five chapters draw upon a common conceptual framework for the analysis of hazard management. This framework, developed at Clark University's Center for Technology, Environment, and Development (CENTED) over the past seven years and elaborated in *Perilous progress: Managing the hazards of technology* (Kates, Hohenemser, and Kasperson 1985), involves three major conceptual models—a model of hazard and related hazard-control opportunities, a model (flow-chart) of hazard management, and a taxonomy of technological hazards. Taken together, the models provide a structure for the systematic analysis of how corporations undertake the management of industrial hazards and how the approaches compare with available options.

To begin, several preliminary clarifications of terminology merit attention. *Hazards* are defined as threats to humans and what they value. *Risks* are the quantitative measures of hazard consequences expressed as conditional probabilities of experiencing harm. *Hazard management* is the purposeful activity by which a corporation informs itself about hazards, decides what to do about them, and implements measures to control them or to mitigate their consequences.

A causal model of hazard. Building upon the customary division of hazards into events and consequences, this model elaborates the evolution of hazards into a series of stages that end in unintended and undesired consequences. The stages flow into each other much like a succession of reservoirs, each leading to the next. Each linkage in this causal chain presents an opportunity for intervention via control measures designed to arrest further evolution of the hazard.

The model begins with an "upstream" component of the hazard in which a basic human need (e.g., energy) is converted into human wants (electricity). Still in the upstream portion of the hazard chain is the choice of technology, involving considerations of realizing benefits and minimizing risks. Thus, in Figure 1.1, the need for energy results in human objectives to use nuclear power to produce electricity. Inevitably, initiating events (e.g., break of a large pipe) can trigger a core melt at the plant, leading to a release of radioactive materials (e.g., radioactive iodine). The "downstream" portion of the hazard chain consists of the exposure of nearby populations to these releases, leading to adverse consequences (in this case, cancer of the thyroid). At all stages, as shown by the bottom row of control interventions, opportunities exist to block the evolution of the hazard. Downstream management options, involving prevention of consequences once exposure has occurred, are not very promising, or even possible. Upstream options involving human wants are possible and have a large impact upon potential hazards. But once the choice of technology occurs and nuclear power is adopted, then "mid-stream" interventions to prevent events and releases become the predominant controls.

10

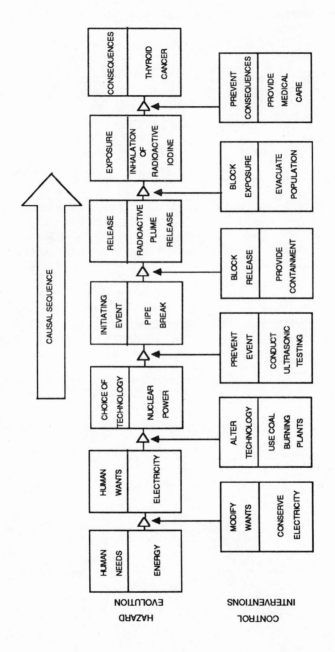

Figure 1.1 The hazard chain model, as applied to a hypothetical nuclear plant accident.

Even this elaborate model offers a very simplified structure of hazard, for feedback occurs among the stages. Yet, the model is useful in that it provides a standardized means for structuring hazards and for identifying systematic opportunities for hazard control. Each stage in the evolution of hazard must be assessed for the scientific understanding of its link to the subsequent stage and for the potential blocks that could arrest the progress of the hazard toward adverse consequences. Not shown is a parallel benefit chain that also results from the choice of technology.

A flow-chart of hazard management. Hazard management, whether by a government agency or a corporation, has two essential functions: intelligence and control. Intelligence provides the information needed to determine whether a problem exists, to define choices, and (retrospectively) to determine whether success has been achieved. The control function consists of the design and implementation of measures aimed at preventing, reducing, or redistributing the hazard, and/or mitigating its consequences.

Figure 1.2 depicts the management process as a loop of activity. In the center of the diagram is the hazard chain through which the deployment of technology may cause harmful consequences for human beings and their communities. Four major managerial activities—hazard assessment, control analysis, strategy selection, and implementation and evaluation—surround the chain. Each, as becomes apparent in the case studies that follow (chapters 2-6), characteristically involves both normative and scientific judgments. The depicted sequence, of course, is an idealization and simplification of a process that is often not linear or which jumps over stages.

Hazard assessment involves four major steps—hazard identification, assignment of priorities, risk estimation, and social evaluation. A *control analysis* judges the tolerability of the risk and evaluates the opportunities for controlling the risk. A cost analysis is, of course, an important part of this stage as well. Third is the *selection of a management strategy*, composed of a management gaol and a package of control interventions that make up the risk reduction and mitigation program and the precise institutional means selected. Finally, the management program requires *implementation and evaluation*, a stage at which many well-designed efforts break down (witness the case of Bhopal in chapter 6).

A taxonomy of technological hazards. The final conceptual model utilized is one that promotes meaningful comparisons of diverse technological hazards. The causal chain model has facilitated the identification of 12 biophysical attributes of hazards the scoring of 93 different technological hazards in terms of these attributes. Eventually, factor analysis reduced the 12 attributes to five, termed *BIOCIDAL, DELAY, CATASTROPHIC, MORTALITY,* and *GLOBAL,* which permit the construction of a seven-class taxonomy (Hohenemser, Kates, and Slovic 1983). These results provide a systematic context for hazard comparison. Since most large corporations, as the following chapters attest, must manage large and variegated hazard domains, all face the common problem of comparing and setting priorities for hazards. The taxonomy (Table 1.3) informs assessment of the judgments and choices that have been made.

These models receive short shrift here, but fuller treatment is readily available elsewhere (Kates, Hohenemser, and Kasperson 1985). The models must be

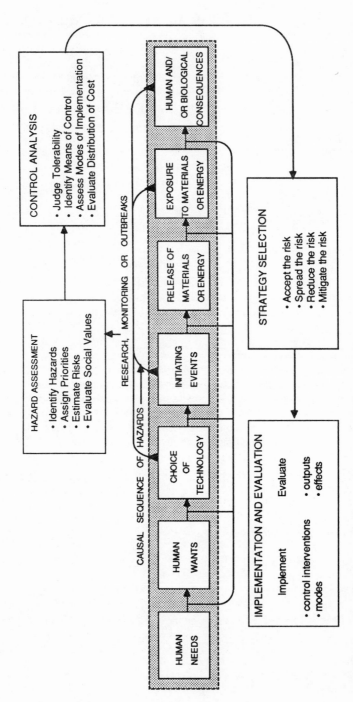

Figure 1.2 A flow chart of hazard management.

Table 1.3

A SEVEN-CLASS TAXONOMY

CLASS	EXAMPLES
Multiple extreme hazards	Nuclear war (radiation), recombinant DNA, pesticides
Extreme hazards 　Intentional biocides 　Persistent teratogens 　Rare catastrophes 　Common killers 　Diffuse global threats	 Chain saws, antibiotics, vaccines Uranium mining, rubber manufacture LNG explosions, commercial aviation (crashes) Auto crashes, coal mining (black lung) Fossil fuel (CO_2 release), SST (ozone depletion)
Hazards	Saccharin, aspirin, appliances, skateboards, bicycles

used flexibly, of course, and in conjunction with other theoretical frameworks and perspectives necessary to clarify the workings of corporations in advanced industrial society. But they do provide a common basis for analysis and facilitate comparisons among the case studies.

Organization

　　The next five chapters characterize the hazard-management programs of five large corporations. They vary, however, in scope of treatment and locus of analysis. Chapter 2 examines one of the large U.S. petrochemical companies for its overall approach to hazard management, including both product safety and occupational health protection. Chapter 3 focuses on the management of occupational hazards at a specific plant of a large pharmaceutical company. Chapter 4 switches to the international level to analyze the product-hazard program of a Swedish automobile manufacturer. The fifth chapter examines the defense-in-depth hazard management program at a government-financed hazardous facility. Chapter 6 also focuses at the international level—in the case an analysis of the failure of industrial hazard management and the resulting Bhopal tragedy. Finally, chapter 7 draws together the major generalizations and insights emerging from the case studies, and assesses their implications for public policy and research needs.

2

Managing Hazards
at PETROCHEM Corporation

PETROCHEM Corporation, a fictional name to provide anonymity, is one of the largest petrochemical companies in the United States, ranking in the top 10 percent of the Fortune 500. Annual corporate revenues exceed $20 billion, with approximately 75 percent of capital and exploration spending on the development of domestic energy resources. PETROCHEM has over 30,000 employees, distributed over 15 manufacturing locations, including chemical plants and refineries plus exploration, field, laboratory, and office workers. The corporation remains primarily an oil and gas company; its chemical division employs fewer than one-fourth of all employees and accounts for less than one-sixth of company sales. It does, however, deal with a broad range (approximately 1,500) of individual chemical products.

This chapter inquires into hazard management at PETROCHEM Corporation. The goal of this effort is to characterize in some detail how this major corporation, one of the leaders in industrial health and safety, organizes its hazard management programs as they pertain to two major classes of hazards—those associated with product safety and those that involve protection of workers. In such an exploratory study, the researchers are not in a position to reach judgments as to how "good" or "bad" hazard management at PETROCHEM is although evaluative comments are provided where warranted by available data. Similarly, the research does not attempt to explain the motivations that lay behind individual or corporate decisions. Data bases for the study include extensive documentation (trend data, risk manuals, policy statements, internal memoranda) provided (very fully and openly) by PETROCHEM, two lengthy visits to corporate headquarters during which extensive interviewing occurred, and follow-up meetings and discussions with PETROCHEM officials. Throughout, interview results and documentation were compared for consistency and, wherever possible (such as injury trend data), with independent sources.

The Structure of Hazard Management

As a large, profitable corporation, PETROCHEM commands extensive financial and organizational resources that can be allocated to hazard management. Over the past decade, an articulated hazard management structure, including specialized expertise and risk evaluation resources, specified health and safety objectives and decision-making procedures, an internal standard-setting and audit system, and close links with major industrial trade organizations, has emerged. A brief overview of that history precedes a detailed discussion of the major features of the management system.

A historical note. A well-developed hazard management system at PETROCHEM has not always existed. PETROCHEM was slow to embrace industrial hygiene. Indeed, in 1971, the company could claim only a very limited industrial hygiene program, little or no internal corporate medical capability (it was entirely by contract), and scant focus by senior management on hazard assessment. Health and safety functions were fragmented throughout the corporate organizational structure. This limited focus and capability exacted their price: with an occupational injury rate that lagged behind a number of competitors in the industry, PETROCHEM was obviously not well positioned to respond to the major federal health and safety legislation and regulation that emerged during the early 1970s.

During 1971 and 1972, PETROCHEM initiated actions to redress this shortcoming. The company appointed a general manager charged with the coordination of health and safety responsibilities and recruited a scientist with outstanding credentials in occupational health protection to establish an industrial hygiene program. Also, the recruitment of a corporate medical staff served to augment the existing program of contractual medical services.

In the mid-1970s, PETROCHEM took a second major step, one that erected the basic outlines of the present management system. A major reorganization integrated all corporate health, safety, and environment functions in one separate administrative unit, headed by a vice president. The latter step afforded health, safety, and environmental protection with access to top corporate decision making. The recognition that in-house capability was essential brought the establishment of a toxicology laboratory, one of approximately 15 such major industrial facilities in the United States (prior to this, external laboratories had performed toxicological work). A series of management innovations—an expanded internal audit system for environmental protection (described below), a systematic health-and-safety review of every product line, a series of task forces to examine in depth particular hazards of concern—quickly followed.

Since the reorganization, hazard management at PETROCHEM has grown in scope and elaboration. Audit systems for occupational health and safety and for environmental protection have been added and a third for product safety has recently been completed. The formulation of a conceptual risk assessment framework guides PETROCHEM's response to individual hazards. Through March, 1985, the company had enacted 11 internal standards to limit worker exposure to hazardous substances, and another nine were under consideration. PETROCHEM establishes such standards in cases where legal standards do not exist or where legal standards are considered to be too high (permissive). The company has also created a computerized health surveillance system that integrates various exposure and health data. A risk evaluation group, instituted in 1985, coordinates corporate risk assessment and seeks to keep the corporation abreast of scientific developments in this field.

Goals and objectives. PETROCHEM uses a formal goal- and objective-setting process to direct its hazard-management program. At its most general level, a corporate statement of philosophy (Figure 2.1) guides the approach to health and safety. This statement of philosophy is notable in (1) its designation of health, safety, and environmental protection as a *primary* consideration in corporate planning and decision making, (2) its statement of a goal to make PETROCHEM a "recognized leader" within the industry in health and safety, and (3) its inclusion of a clear statement of accountability by management and employees alike. These overall objectives are shared by PETROCHEM's various divisions, including the Research and Development Unit, which also recognizes a goal "to become a sustained leader in safety performance among the R & D organizations of the major oil companies" (Internal memorandum, 26 July 1982).

General corporate aims translate into specific objectives for assessing health, safety, and environmental performance. PETROCHEM's occupational safety objectives include: (1) decreasing the number of worker recordable incidents (fatalities and accidents involving days lost from work) and injuries by at least 8 percent per year, and (2) eliminating all industrial fatalities (of PETROCHEM employees). In addition, each unit is responsible for the adequacy of its health, safety, and environment programs. All divisions and, in turn, individual plants and departments, have individual objectives that are parts of PETROCHEM's overall performance. Specific activities are identified as means for achieving annual objectives.

Informal objectives also exist for the hazard managers. Within the corporation, success and "leverage" depend upon the ability to show that an aggressive hazard-management program is *income protective*, if not *income productive*. A long-term view of corporate success, as well as of cost-effective approaches to hazard control, is essential in this respect. Externally, PETROCHEM's hazard managers see themselves in competition with their peers—their counterparts at other large petrochemical companies—and professional standing rests in part on corporate program development and relative performance on the few standardized indicators (e.g., mortality, injury rate, and time lost data) that exist.

Resources and capability. The resources that PETROCHEM can bring to bear on these goals are substantial. The health and safety unit at corporate headquarters numbers approximately 120, including such disciplines as chemistry, biology, engineering, public health sciences, industrial hygiene, medicine, biostatistics, epidemiology, and regulatory expertise (with technical support from other units). Within the corporation as a whole, 30 industrial hygienists serve the population of employees. The toxicology laboratory employs around 50 people, including 20 toxicologists, and the headquarters toxicology staff, administered by the toxicology laboratory, includes three product-process toxicologists. Each major manufacturing location (i.e., plant, refinery) employs at least

Figure 2.1

PETROCHEM CORPORATION
HEALTH, SAFETY,
AND ENVIRONMENTAL CONSERVATION

Statement of Philosophy

At PETROCHEM the health and safety of our employees and protection of the environment must be primary consideration as plans are made and before actions are taken. We believe that excellent performance in health, safety and environmental conservation is essential to the current and long-range success of our business and contributes significantly to good employee and community relations.

In health, safety and environmental conservation, the performance objective of the organization is to be a recognized leader. We intend that our facilities be operated on the basis that virtually all accidents can be prevented and that unacceptable health effects and environmental impacts can be appropriately controlled.

Management has the responsibility, and will be held accountable, for preventing such incidents. We expect managers to fully accept their responsibility for contributing to this objective by exerting appropriate emphasis, establishing and maintaining a committed organization, training employees, providing and enforcing good work procedures, providing and maintaining equipment that is safe to use, and performing ongoing audits of the activities under their control.

Each individual employee also has the responsibility to protect his own safety and health, and to help protect the environment and the safety and health of co-workers by performing work in a manner consistent with established procedures and safe work practices.

one part-time physician and a nurse (two of the largest plants have their own full-time physicians) and at least one industrial hygienist or industrial hygiene technician (larger facilities may have several).

These resources are supplemented by experts brought in to perform contractual work (as in epidemiology), by special training of other PETROCHEM professionals to assist in hazard monitoring and identification (as with corporate sales personnel in monitoring product safety), and by close links with hazard research programs in the major trade associations (as with the American Petroleum Institute and the Chemical Manufacturers Association). The ties with trade associations add very significantly to risk-assessment capability—a PETROCHEM toxicologist estimated the value at 50-100 percent of the company's current budget for conducting requisite toxicological studies. These resources account for a significant, although unknown, portion of PETROCHEM's total budget (one estimate for environmental protection activities for refineries was 10-15 percent of capital investment). The largest proportion of hazard-management effort (estimated by PETROCHEM officials at about 60-65 percent) is allocated to the chemicals division (and particularly to pesticides) of PETROCHEM.

Despite the scale of these resources, PETROCHEM hazard managers acknowledge that most of their effort over the past five years has been reactive (to new legislation, to hazard events). But they also maintain that they are moving to a situation that will be more proactive, particularly in regard to research in hazard identification and risk assessment. The new risk assessment group, the expanded toxicology program, the addition of a biostatistician and an epidemiologist to the staff, the computerized health-surveillance system, and an all-cause, three-year epidemiologic study of one of its larger plants are all examples of the changing orientation in health and safety at PETROCHEM.

Standard-setting and compliance. PETROCHEM's internal regulatory system for hazard control closely parallels that of the public sector—with many of the same functions, roles, processes, and conflicts to accomplish its corporate hazard management goals. PETROCHEM complies with the various governmental (e.g., EPA, OSHA, FDA) standards, sets standards of its own, augments both types of standards with more informal and location-specific target exposure levels, employs a formal structure for judging internal risk standards (and implicitly, priorities), utilizes elaborate deliberative means of conflict resolution, and seeks to assure implementation through formal audit and compliance review and a system of related management incentives.

The first, and major, layer of hazard regulations at PETROCHEM originates in the external environment, particularly in the standards of federal governmental agencies but including state regulations and consensus industry standards and threshold limit values (TLVs) as well. The position of PETROCHEM hazard managers is to contest vigorously any proposed law or standard-setting activity in the legislative or decision-making process if the proposal is judged unreasonable or based upon faulty (or, at least, ambiguous) scientific evidence, data, or interpretation. PETROCHEM often prefers to register its opposition through trade associations, thereby assuring coordination with other industry members, reducing process costs, and avoiding public stigma of seeming opposition to health and environmental protection. In some cases, where PETROCHEM has confidence in other corporations that are actively involved (perhaps because of a major stake in the product under consideration) in a particular case of risk regulation, it may allow them to "carry the industry flag." Once a law or standard is enacted, however, it is adopted as a regulatory requirement and vigorously implemented within PETROCHEM.

Beyond these governmental standards, PETROCHEM adopts internal standards that exceed governmental standards, increase their stringency (by as much as a factor of 10-20), or, if PETROCHEM determines the necessity of additional stringency, institutes a standard even where none exists in federal or state regulation. Such standards have been enacted only in occupational health protection and represent the exposure level below which PETROCHEM expects to operate 95 percent of the time. Environmental protection and product safety have not to date been the subject of internal standard-setting. PETROCHEM has adopted 11 internal standards (Table 2.1), with another nine in process. Such standards, although internal, are considered to have a force nearly that of law in that individuals could conceivably invoke them in court cases against PETROCHEM. In addition, stringent internal corporate standards may be part of a justification or argument for the feasibility of a proposed federal or state regulation.

Given these risks, one might ask why PETROCHEM, or any other major industrial corporation, would institute internal hazard standards. Company officials cite four motivations: (1) to reduce health effects to workers (and thereby make for a safer workplace), (2) to protect PETROCHEM against product liability claims, (3) to anticipate regulation, thereby ensuring that upgrading can be accomplished in an orderly and cost-effective way, and (4) in cases where PETROCHEM already has very low exposure levels, to achieve a competitive advantage in encouraging a more stringent federal government standard. In regard to the last, this motivation is, according to PETROCHEM officials, pursued only sparingly by corporations because of a general industry aversion to encouraging regulation and because the next hazard case may find the competitor with the comparative advantage in hazard reduction. Other industries take note and may themselves reconsider their internal standard (particularly if they have confidence in the risk assessment program of the corporation), partly because of greater vulnerability in product liability and also for reasons of health and safety.

Guidelines for supplementing external and internal standards comprise PETROCHEM's "local operating targets" for locations where exposure assessments and operating considerations suggest that it is both desirable and feasible to operate below the limits mandated by legal standards. These targets are expected to be attainable in the near term. They do not constitute an upper limit for safe operation and, if exceeded, do not always prompt immediate corrective action.

The process for setting an internal PETROCHEM standard is formal, time-consuming, and designed to embrace the variety of considerations and viewpoints that exist within the corporation. Once a standard-setting process is initiated within the health, safety, and environment unit in corporate headquarters, a specially appointed task force sets out to gather the best available health science information as well as other data relevant to impacts on the corporation. These "other" considerations include such issues as cost, impact on operations, required employee practices, anticipated external regulatory action (if any), cost-effective engineering, and design and construction alternatives. Intuitive judgments are allowed but they must be explicit. A broad range of consultation within the corporation, including legal staff and (where necessary) representatives from production plants to be affected, occurs. Resolution of inevitable conflicts occurs through negotiation and accommodation among interested parties, who formulate a specific recommended standard, with supporting justification. The draft standard then undergoes wide review within PETROCHEM and eventually requires sign-off by the directors of corporate medical services, safety and industrial hygiene, and the toxicological laboratory as well as a consultant in biomedical sciences, and (eventually) by the Vice President for Health, Safety, and Environment. Characteristically, this corporate "rule-making" process will have consumed several months of effort.

The internal regulatory system requires judgments about the levels of risk that require attention versus those that are so low as to be relatively insignificant. PETROCHEM utilizes a formal system of evaluation that is conceptually based upon three risk regions; high risk, low risk, and insignificant risk (Figure 2.2). The high-risk region includes those risks judged to require immediate corporate action in order to meet the objective of risk reduction. Risks in the low-risk region rank below the level that triggers immediate response and may be said to be variously acceptable, depending upon the particular circumstances. They are viewed as representing additional opportunities for overall risk reduction. Risks in the insignificant-risk region are judged not to require corporate risk reduction actions since reduction would not add significantly to health benefits.

The primary operational focus is on the boundary line—the *level of action*—between the high-risk and low-risk regions. The level of action represents the functional trigger for risk-reducing measures. For cancer the level of action is set at a lifetime (30-year) risk of 2×10^{-3}, which is equivalent to an annual risk of 1 in 15,000. The lifetime risk of 2×10^{-3} corresponds to PETROCHEM's judgment of .1 percent of the risk of developing cancer from all causes, including diet, smoking, lifestyle, etc. The risk level is also derived from comparison with various extant risks in society and from defining work as a semivoluntary activity. A secondary focus is on the boundary line—the *level of insignificance*—between the low-risk and insignificant-risk regions. For cancer, this level is typically set at 10^{-5} (or at the low-risk level).

The setting of levels of action and levels of insignificance for other biological effects follows the principle of risk reduction. Typically the level of action is set at some fraction of the background rate for the comparable health effect when that can be determined with some degree of confidence. The nature and severity of the health effect influence the magnitude of the difference between the level of action and the background rate.

After judging risk level and recognizing standards or local operating targets, the corporation must assure compliance by its various plants, refineries, divisions, and departments. PETROCHEM employs several means to do this. An internal audit system, the centerpiece of compliance assurance, exists for

Table 2.1

OSHA PERMISSIBLE EXPOSURE LIMITS (PELs) AND PETROCHEM INTERNAL STANDARDS
(through March, 1985)

MATERIAL	OSHA PEL	DATE	PETROCHEM STANDARD	ACTION
Epichlorohydrin (ECH)	5 ppm (8-hr. TWA)	12/77	1 ppm (8-hr. TWA) 3 ppm (15-min. Peak)	Workplace monitoring, engineering controls, work practices, or personal protective equipment, plus medical surveillance
Benzene	10 ppm (8-hr. TWA)	2/79	Same as OSHA	Same as for ECH, above
Ethylene Oxide	50 ppm (8-hr. TWA) (a)	10/78	5 ppm (8-12 hr. TWA) (a) 15 ppm (15-min. Peak)	Same as for ECH, above
Dichloropropane dichloropropene mixture	None	7/76	1 ppm (8-hr. TWA)	Workplace monitoring, engineering controls, work practices, or personal protective equipment
Ethylene Dibromide	20 ppm (8-hr. TWA) 30 ppm (ceiling)	3/78	1 ppm (8-hr. TWA)	Same as for dichloropropane, above

Alpha Olefins, C$_6$-C$_{18}$	None	1/78	100 ppm (8-hr. TWA)	---
Carbon Tetrachloride	10 ppm (8-hr. TWA)	8/82	5 ppm (8-12 hr. TWA)	Same as for dichloropropane, above
Ethylene Glycol Monomethyl Ether	25 ppm (8-hr. TWA)	6/84	2 ppm (8-12 hr. TWA) 6 ppm (15-min. Peak)	Same as for dichloropropane, above
Ethylene Glycol Mono-ethyl Ether Acetate	100 ppm (8-hr. TWA)	8/84	5 ppm (8-12 hr. TWA) 15 ppm (15-min. Peak)	Same as for dichloropropane, above
Carbon Disulfide	20 ppm (8-hr. TWA) 30 ppm (ceiling)	11/84	5 ppm (8-12 hr. TWA) 30 ppm (15-min. Peak)	Same as for dichloropropane, above

(a) On August 21, 1984, OSHA published a PEL of 1 ppm ethylene oxide (8-hour TWA). OSHA did not, at that time, establish a Short Time Exposure Limit. PETROCHEM, however, continued a 15 ppm (15-minute) Peak Exposure Limit for their facilities as a control mechanism or aid in keeping exposures below the 1 ppm 8-hour TWA.

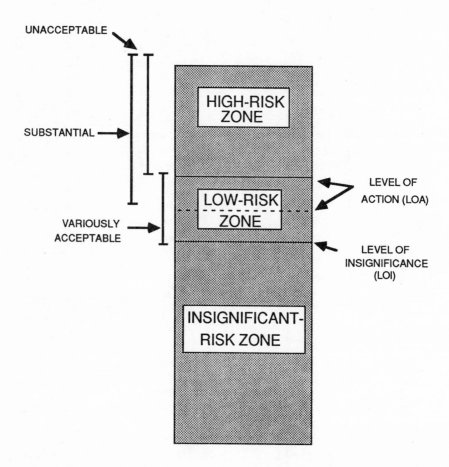

**Figure 2.2 PETROCHEM'S risk-region approach
to judging risk acceptability.**

environmental protection, occupational health and safety, and product-safety functions. The objective of such reviews, as stated explicitly by PETROCHEM, is "to provide assurance that systems for verifying that laws, regulations, and policies pertaining to safety and health exist and are working effectively in all the functions" (see below). A multivolume resource manual spells out objectives, review criteria, external and internal standards, PETROCHEM accident-investigation procedures, etc.

Corporate headquarters audits principal PETROCHEM locations every three years (or more often in cases of special need), with audits covering (sometimes in separate processes) environmental protection and safety, health, and product safety. The scope of the audit depends upon the facility under review; small facilities may require only one auditor to cover both health and safety, whereas large facilities may require an auditing team. Such a team will typically be drawn primarily from outside the facility and will include four or five persons— one health and safety professional, one engineer, and one or two persons with health and safety responsibilities at other PETROCHEM locations. A health-and-safety official from corporate headquarters is the team leader on all major audits and one other headquarters member sits in on all wrap-up sessions. Although representatives from the corporate medical division do not normally participate on the team, they occasionally accompany review members to assess the quality of medical programs. Occasionally an attorney participates.

The compliance assessment ranges in duration from three days for a small facility to as much as two weeks for a large manufacturing plant. At the conclusion of its work, the audit team prepares a written report of 5-25 single-spaced pages, which is reviewed with the location manager and provided to health and safety officials at corporate headquarters. The process emphasizes cooperation and seeks "to have everyone buy into the report." All serious deficiencies are brought to attention of the health and safety managers and followed up until the situation is corrected. Minor issues (the bulk of cases) are left to the discretion of the location manager for rectification but future reviews track overall performance. Since evaluations of the location manager take into account performance in the area of health, safety, and environmental protection as well as productivity, incentive exists to institute the recommended improvements.

These compliance reviews comprise a significant activity in PETROCHEM's health, safety, and environment organization. From its inception in 1981 the program had conducted a total of 321 reviews through 1984. PETROCHEM officials report the discovery and rectification of a few serious deficiencies, characterize the bulk of deficiencies as "minor," and point out that even these minor shortcomings are slated for corrective action.

The Vice President for Health, Safety, and Environment receives quarterly reports that summarize overall PETROCHEM performance, and, in turn, PETROCHEM's top management receives a summary to ensure that the recommendations in an audit have been implemented.

Matrix management. The hazard-management process at PETROCHEM shares many characteristics of the overall management structure of the corporation. PETROCHEM uses many elements of a matrix system of management in which a number of functions (including Health, Safety and Environment) have ties to two product organizations (Oil and Chemicals) and also coordinate their activities with certain other organizations (e.g., exploration, production activities, and subsidiaries) and other functions (e.g., R & D, marketing, and manufacturing). Figure 2.3 provides a conceptual sketch of this system at PETROCHEM, both in its general form and in its relationship between corporate headquarters and individual plants.

This management structure functions both as an intelligence and a decision system. Monitoring of existing programs and proposals for new actions require information gathering, consensus-building, and (where conflict occurs) bargaining and negotiation. The company places a premium upon broad consultation through the matrix and the attainment of a high degree of consensus. As a result, management decision making tends to be cumbersome, slow, and expensive. At the same time, such a management system minimizes the likelihood of miscalculation and error and eases problems in implementation.

The formal organizational structure of PETROCHEM's Health, Safety, and Environment division is relatively flat (Figure 2.4). The characteristic vehicle for shaping management responses on all important issues is decision evolution by committee deliberation. PETROCHEM hazard management is rich in committees and task forces. Table 2.2 provides a summary of the four major standing committees and their functions. These committees sometimes overlap in function or jurisdiction, but such problems are sorted out in practice. The committees are central to the formulation of major policies and programs and are not subject to by-passing by the formal administrative structure (except for more routine matters). Even for specific tasks, the tendency is to assign them to special work groups, which often become broad-based working committees, rather than to specific individuals. All critical proposals move up to the Vice President through task groups and committees that serve dual functions of scientific validation and political mediation. In the great majority of cases, by the time the Vice President makes a decision, extensive deliberation has occurred, issues have been fully aired, and substantial consensus achieved within (as well as outside) Health, Safety, and Environment.

The foregoing overview of hazard management at PETROCHEM sets the context for the details of product-safety and occupational-hazard management.

Product-Safety Management

The range and nature of products sold by PETROCHEM Corporation affect substantially its hazard-management program. Since the mid-1960s, PETROCHEM has, in addition to its oil and gas products, acted primarily as a producer of intermediate chemicals for heavy-volume, sophisticated users. In fact, only two of its locations manufacture chemical products (pesticides, antifreeze) that go directly to the consumer. As a result, PETROCHEM has not emphasized the problems involved with chemical end uses. Nonetheless, with some 1,500 individual chemical products, PETROCHEM must evaluate and respond to a sizeable range of potential chemical hazards, an activity that occupies approximately 80 percent of the effort of the product-safety unit. PETROCHEM has no formal product stewardship program, it should be noted, although it does have some

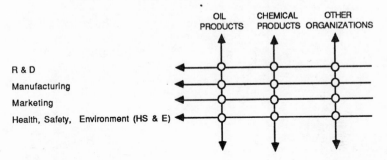

GENERAL MANAGEMENT STRUCTURE

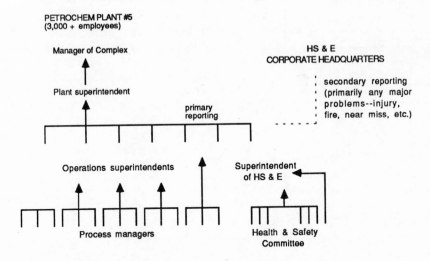

CORPORATE HEADQUARTERS/PLANT RELATIONS IN HS & E

Figure 2.3 Matrix management at PETROCHEM Corporation: Schematic diagrams.

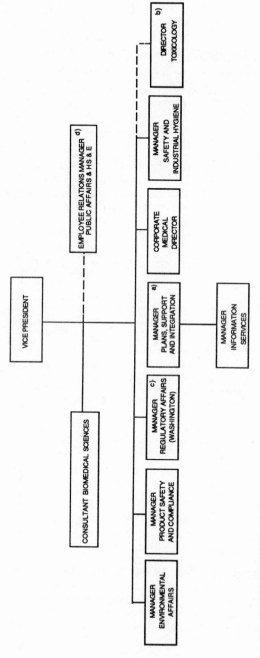

a) Serves as deputy to Vice President
b) Reports administratively to PETROCHEM Development and functionally to HS & E
c) Reports administratively to Public Affairs (Washington) and functionally to HS & E for HS & E matters
d) Reports administratively to Vice President Public Affairs and functionally to Vice President HS & E for HS & E matters

Figure 2.4 PETROCHEM Corporation: Health, Safety, and Environment (HS & E) Organization.

Table 2.2

PETROCHEM CORPORATION

**MAJOR COMMITTEES IN HEALTH,
SAFETY, AND ENVIRONMENT**

Health, Safety, and Environment Group
Established in late 1984, this is the toxicology issue and management oversight body. It functions in areas where the hazard issue transcends any single department or group. Its functions include: advice in priority setting, identifying decision options, resolution of organizational conflicts, implementation strategies, and review of past decisions. Chaired by the Vice President, it is a high-level committee for broad-based analysis of major hazard issues.

Senior Safety Steering Group
This committee includes the senior staff members of the division, including representatives from major subsidiaries. it serves primarily coordination and communication functions and meets approximately every six weeks. It also provides an ongoing technical link between Health, Safety, and Environment and other corporate functions on workplace health and safety issues. It is one important source for the formulation of internal regulations.

Environmental Steering Team
Chaired by the Manager of Environmental Affairs, this group includes representation at the manager level of the various units (including Pipe Line and the PETROCHEM Development Company) which are involved in environmental questions. Particularly this group acts as a forum for proposed environmental regulations and legislation. It meets monthly.

Risk Evaluation Group
Established in 1983 as the risk assessment arm of Health, Safety, and Environment, this group is composed of 10 members drawn from skills central to risk assessment issues. It was formed particularly to assume a key role in assessing product risks. The group is responsible for ensuring that risk evaluations have included all necessary steps, for providing documentation, and for identifying options for risk reduction. Members also assist in assigning priorities and act as "tutors" for the rest of the staff in risk assessment developments. At this writing, the group was systematically reviewing PETROCHEM's 20 existing or proposed internal standards.

elements that would go into such a program. It does have a procedure for informing primary customers on safety, health, and environmental matters.

Hazard identification and assessment. PETROCHEM officials view as a continuing process the evaluation of their products for health and safety effects. A Product Review Task Group, established to review the company's products on a systematic basis and to search for unforeseen hazards, acts as a driving force for further investigation and also identifies products for additional health-effects studies. A second driving force for the development of health-effects information on products is the Toxic Substances Control Act (TSCA), which requires the submission of a Pre-Manufacturing Notification, containing information on known health effects, to the Environmental Protection Agency at least ninety days before the manufacture of a new chemical compound or substance for commercial purposes.

As new health-effects information on specific products is developed (internally or externally), PETROCHEM reviews the data to determine whether a product warrants new or additional handling precautions, user information, or labeling. The determination frequently follows a study of the available health and safety data by the PETROCHEM Risk Evaluation Group. Regulations promulgated under TSCA, the Federal Insecticide Fungicide and Rodenticide Act (FIFRA), and the Consumer Product Safety Act also contain requirements for the reporting of substantial (previously not identified) risks.

If the presence of a hazard is suspected in an existing product (possibly from information provided by a downstream user) or in one under development (usually by corporate assessment procedure), a Hazard Evaluation Group, chaired by the Manager of Product Safety and Compliance and including representatives from Toxicology, Environmental Affairs, Plans & Support, and Industrial Hygiene (and others as needed), gathers and assesses all relevant information. The group then provides a formal interpretation of its findings to the Vice President who must decide whether the information is reportable within the meaning of one of the three acts. Preparing this assessment involves significant effort— several weeks of effort by perhaps five or six persons, entailing some 100-120 person-hours of work. The hazard Evaluation Group may also identify the possible need for hazard-control actions that can be undertaken by PETROCHEM. These are referred to the appropriate organization for action.

New products receive special attention by PETROCHEM early in the development stage. If the product is a chemical substance not previously manufactured for commercial use, then a Pre-Manufacturing Notification (PMN) must be submitted to the EPA. In addition to this, however, health and safety officials recognize a need to advise the business side of PETROCHEM about perceived labeling and safe handling requirements that could affect the marketability (acceptance) of the product. In this connection, typical questions often include: Will it be a consumer product? What will be the level of exposure? Are children likely to be exposed? What are potential misuses of the product?

PETROCHEM product-safety personnel believe that they have succeeded in dealing with the acute hazards of PETROCHEM products (a substantial portion of which involve unintended uses). Chronic hazards, by contrast, are a large and growing problem. Identification of a potentially serious chronic hazard in a new product usually generates at PETROCHEM a reluctance to market it unless the hazard issues can be resolved. If the product's profitability is judged unlikely to be capable of sustaining the necessary toxicology cost, PETROCHEM will abandon the development. If, on the other hand, it promises to be a large-volume product with a substantial market, a lengthy evaluative process and assessment of potential options for hazard control may produce a decision to "run the regulatory risks" inherent in moving ahead with the product. This decision is a broad-based one, involving both the business and the health, safety, and environment components.

Pesticides are a notable example of new-product hazard management. On the chemical side of PETROCHEM's business, agricultural chemicals provide a substantial part of profits but represent a relatively small percentage of investment. Examples exist where PETROCHEM has gone through all stages (including building a new plant) in the development of a pesticide and failed to pass federal requirements for registration. Since it can take approximately $10 million and eight years of testing (for chronic hazards) to develop a new pesticide, substantial economic risk is at stake. PETROCHEM takes the position, based on experience, that a certain number of failures will inevitably occur but that the potential return may merit the economic risks for a promising product. Such a developmental effort heavily involves the product-safety representatives.

Monitoring of products provides another means of hazard identification. PETROCHEM salespersons, as required by TSCA and the Consumer Product Safety Act, must bear the responsibility for reporting suspect hazards. But, of course, this method uncovers few chemicals that pose chronic hazards (and then they usually are flagrant), for most chronic-hazard identification surfaces through the toxicological laboratory and through review of the literature.

PETROCHEM employs a formal ongoing product survey of health information, designed over a 5-7 year period to identify major potential health issues for all PETROCHEM products. A random product-selection process identifies the 300 products to be reviewed each year. To be eligible, the candidate product must be currently marketed by PETROCHEM, have Material Safety Data Sheets, and must not already be under review in other PETROCHEM high-priority analyses. An initial search of the published and proprietary literature for each product determines the relevant data base. A selection of major references and review articles comprises a reference list (with abstracts) for consideration by an interdisciplinary PETROCHEM review group. Members of the review group evaluate the citations in their areas of expertise and prepare summaries of key data and health issues that need to be addressed. These summaries, in turn, proceed to an overall Product Review Task Group, which reviews all the data summaries, summarizes the main hazard issues, establishes priorities for further PETROCHEM action, assessment, or testing, documents the rationale for its recommendation, and submits the analysis for permanent storage and retrieval in a special PETROCHEM data base.

Hazard control. Given its situation as a largely bulk producer of intermediate chemicals, it is not surprising that PETROCHEM emphasizes informational strategies in hazard minimization once design and handling safeguards have been included. At the most basic level, this involves extensive product labelling for hazards and wide dissemination of the federally mandated material safety data sheets (MSDSs), related safety brochures, safety bulletins, and more extended technical bulletins for various PETROCHEM chemical products. In addition, PETROCHEM has developed an overall *Chemical safety guide* and disseminated some 500,000 copies to distributors, customers, poison control centers, regulators, etc. All this literature undergoes extensive review within PETROCHEM, particularly by the toxicologists and medical personnel.

Assuring compliance by "downstream" users is inherently difficult for any large corporation. But once aware of a potential problem at a distributor or user (usually a small company), PETROCHEM, at the request or with the agreement of the distributor customer, dispatches industrial hygienists to help on protective measures. PETROCHEM may also offer assistance for the user to upgrade hazard control. Should both approaches fail, PETROCHEM, if it views the problem as sufficiently serious, may even stop selling to the company (an action, according to PETROCHEM officials, that has occurred on several occasions).

Liability. Product liability is, in the words of one manager, a "hell of a problem" for PETROCHEM. The intrinsic difficulty is that, as an intermediate producer, PETROCHEM is usually a second, third, or fourth party in a product-liability suit. At work is what PETROCHEM product-safety managers brand the "deep pocket" syndrome—suits that attempt to reach back a number of steps in the chemical products manufacturing and distribution chain to locate the party with the "deepest pocket." Juries, viewing an injured party and a large petro-chemical company, may make substantial awards whatever the evidence. Although annual paid product-liability losses for PETROCHEM showed relatively little change between 1978 and 1981, since then the corporation's losses have increased substantially. Concerned that this may represent a substantially growing problem on the horizon, PETROCHEM officials acknowledge that product-liability concerns have stimulated greater attention to product-safety programs. They are also producing a much greater focus upon end-use hazard assessment of chemicals.

Occupational Hazard Management

As noted earlier, substantial development has occurred in PETROCHEM's occupational safety and health programs. Directing these is a formal system of goals and objectives that seek to place PETROCHEM among the health and safety leaders of the industry. Within this context, the discussion to follow examines how these programs bear upon hazard identification and assessment, hazard control, and the working of the management system.

Hazard identification and assessment. This subject calls for discussion in the context of worker safety (acute hazards) and worker health (chronic hazards). PETROCHEM gives attention to both types of hazards.

PETROCHEM has a well-developed system for monitoring occupational accidents and the more significant near-miss accidents. A standardized format for accident investigation and reporting categorizes accidents and defines the scope of the investigation required and the composition of the investigation team. A required formal report identifies the causes and needed corrective actions of the accident or near-miss. In addition, each facility is responsible for regularly providing standard data on occupational injuries and fatalities.

Identification and assessment of health hazards are a much more difficult matter and PETROCHEM hazard managers acknowledge that they are only now moving to a strong identification capability. To this end, PETROCHEM has created a computerized health-surveillance system (Figure 2.5) that comprises six data bases:

- **demographic data** derived from standard personnel and payroll systems and covering all pensioners, active employees, and workers terminated since 1970;
- **biometrical data** covering medical examination data such as blood pressure and pulse, blood chemistry and urinalysis, medical history, smoking history, and clinical diagnoses;
- **morbidity data** from the corporate medical departments;
- **mortality data** from death certificates for pensioners and for employees who die while on active duty;
- **work histories**, including location of work, description of work performed, and chemical and physical agents on the job; and
- **exposure data** from various monitoring and sampling programs.

The system was implemented at PETROCHEM's 15 manufacturing locations in 1979 and at distribution plants and exploration and production sites in 1981. Systematic retrieval serves to monitor industrial hygiene programs and to identify health hazards that might otherwise not become apparent.

Supplementing the overall health surveillance system is a series of specialized biological monitoring programs. Included is surveillance for such hazards as asbestos (largely involved in asbestos removal), benzene, vinyl chloride, lead, pesticides, epichlorohydrin, and others. Under consideration at time of writing was a new smoking-cessation program for workers.

Recently, PETROCHEM has also sponsored three epidemiological studies to determine whether a suspect hazard posed a health problem. Two of these were initiated primarily because of worker concern in a particular plant; the third, which examined exposure to isopropyl alcohol, was "proactive" in nature and was not a response to any specific concerns. At time of writing PETROCHEM was involved in its most ambitious study—a three-year, $500,000 all-cause epidemiological study of one of its largest plants. University consulting scientists have conducted all of these human health studies, and the results have undergone peer review and have been published in scientific journals.

Hazard control. Once a hazard has been assessed and judged to require reduction (according to the risk-regions approach described above), the various means of control must be evaluated. PETROCHEM hazard managers attach high priority to identifying the most efficient means of implementing control measures for minimizing hazards. If not properly integrated with design and engineering, such actions may prove very costly. In assessing various control options, PETROCHEM hazard managers work very closely with plant managers and production engineers, considering such factors as the work-time regime, the

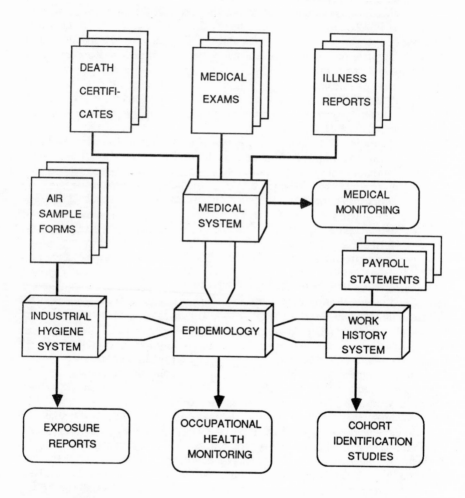

Figure 2.5 The PETROCHEM computerized health-surveillance system.

level of training and knowledge of the employees, the exposure levels, the process operating practices, and the type and nature of equipment involved. Wherever possible, hazard reduction is incorporated into process design, modification, and retrofitting.

PETROCHEM's occupational hazard control program recognizes a hierarchy of preferred control or protection measures. First, preference goes to the use of engineering controls or alteration of work practices to reduce exposure levels. Only if these are infeasible will personal-protection measures be considered, and then the goal will be to restrict their use to a minimum. Throughout, PETROCHEM hazard managers pose the problem to production managers and involve them in identifying solutions.

In addition to the continuing system—of goals, objectives, health and safety upgrading, and compliance assurance—that operates in PETROCHEM management, are programs of worker training and education. These begin in orientation before an employee begins work and continue on a regular basis in monthly safety meetings. These routine practices are supplemented by special informational "alerts" when PETROCHEM adopts a new standard or believes that the hazards of a particular substance require special attention by workers. For other hazards, the education and training program is more extensive—PETROCHEM's 300 asbestos-removal workers, for example, all participate in a slide/tape program designed to indicate the nature of the hazards and the means for reducing exposure.

One distinctive element of PETROCHEM's occupational hazard-control program is a policy on protection of the embryo-fetus, a difficult issue in occupational protection in the 1980s. According to this policy, chemical exposure of employees who are or may become pregnant is scrutinized in order to control the exposure of the embryo-fetus to embryo-fetotoxic and teratogenic substances. The program recognizes three categories of hazard and required response:

Category 1: Job assignments that involve substances that have been suggested to have embryo-fetotoxicity but for which PETROCHEM believes the pattern of evidence does not indicate that the health of an embryo-fetus would be endangered.

Employees in this hazard category are informed of the hazard and if a woman is, or becomes, pregnant, she may request removal from the job. In such a case, she may be reassigned depending upon suitable job availability and existing local employee relations practices and/or labor agreements.

Category 2: Job assignments determined by PETROCHEM as posing a potential threat to the embryo-fetus as a result of cumulative exposure or possible exposure above normal operating conditions but where PETROCHEM believes that threat to the embryo-fetus prior to detection of pregnancy is not significant.

In this case, steps will first be instituted to eliminate the threat to the embryo-fetus. In the interim, females known to be pregnant will be ineligible for these jobs. Existing female workers will be removed from these jobs when they notify their superintendent or PETROCHEM medical staff that they have become pregnant. PETROCHEM then attempts to reassign these employees to other jobs.

Category 3: Job assignments determined by PETROCHEM as posing a clearly defined risk to an embryo-fetus because of the possibility of early embryo-fetotoxic and/or teratogenic effects occurring before a pregnancy is detected.

In such a case (and is should be noted that very few PETROCHEM chemicals are involved), steps are first initiated to eliminate the threat to an embryo-fetus. In the interim all females of childbearing capability, whether they intend to become pregnant or not, whether they practice birth-control, and whatever their marital status, are ineligible for these jobs. PETROCHEM attempts to reassign existing female workers of child-bearing capacity to another job.

An extensive counselling program accompanies the implementation of this policy and information not supporting PETROCHEM's conclusion is made available to the concerned worker. The company has identified some ten of its chemical products as relevant to this policy. Even if no restriction is imposed, all female workers receive pertinent toxicity and exposure information concerning their work assignments. At time of writing, no job reassignments had occurred as a result of the policy, partly because the policy had been anticipated for some time and female workers made aware of these concerns and partly because the number of chemicals involved is small.

Outcomes. What may be said of the overall effectiveness of PETRO-CHEM's occupational health and safety program? The present analysis makes no attempt to provide judgments on how "good" or "bad" certain decisions were. Such an evaluation would go well beyond the intent and capability of an exploratory study aimed at characterizing industrial hazard management and would require reliable data on occupational diseases (such data generally do not now exist). Moreover, one would need to consider trends within industry generally and within the petrochemical industry in particular.

PETROCHEM does, however, employ "yardsticks" by which it measures its own performance (the data for which are available in the public domain). Primary among these is a set of indicators that record—in percentages of workers—death, absence from work, restricted activity, or medical treatment, as compared with the percentage of those with no recordable occupational injuries or illnesses. Figure 2.6 shows the favorable slope of these trends from 1976 to 1983 for the manufacturing activities of PETROCHEM.

More interesting, perhaps, is the total *recordable incidence rate*, expressed in cases per 200,000 hours worked. Figure 2.7 indicates a substantial drop for

37

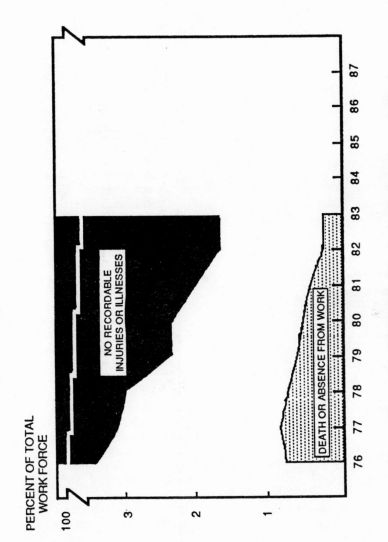

Figure 2.6 Occupational Injury/Illness experience manufacturing.

38

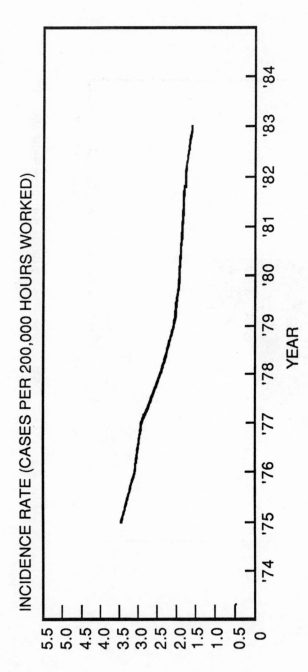

Figure 2.7 Total recordable incidence rates.

PETROCHEM between 1975 and 1983. (For perspective, it should also be noted that the occupational incidence rates for companies reporting to the American Petroleum Institute are about 50 percent the rate of all industries reporting to the National Safety Council and about one-third the rate of the general industry performance reported by the Bureau of Labor Statistics).

PETROCHEM is one of the large, successful petrochemical companies, of course, so that it has greater technical capability in occupational safety and health fields than the average corporation. But even when incidence rates (fatalities plus cases with days away from work) are compared with the industry leaders (and PETROCHEM's major competitors), the trend lines, although not the most impressive of the group, nonetheless again compare very favorably (Figure 2.8). In short, it is apparent that the substantial investment in health and safety and the institution of a wide array of hazard reduction programs point to major improvements on these measures for which data exist.

Conclusions

This case study of PETROCHEM yields a number of tentative conclusions:

1. Beginning from a position of relative weakness some 15 years ago, PETROCHEM has markedly improved its hazard-management programs and capability, characterized by a high level of scientific expertise, strong financial and management resources, and access to the upper echelons of corporate decision making. Contributing factors to the decision to institute a more ambitious hazard-management capability were the need to deal with the array of environmental and health legislation of the early 1970s, a recognition of the company's poor comparative performance in occupational safety, and increasingly activist union interest in occupational protection.

2. PETROCHEM has an elaborate hazard-control structure with roles, deliberative processes, formal assessment, standard-setting, and compliance means that parallel those of the public sector. Conflicts parallel those occurring in rule-making in the public sector and substantial effort is given to conflict resolution and accommodation in shaping corporate health and safety policies. When these processes fail, hierarchical roles and authority become more central in decision making.

3. Corporate health and safety programs are driven partly by external forces such as regulatory requirements and product-liability claims, partly by internal corporate objectives for providing a safe and healthy workplace for its employees, and partly for protection of long-run corporate income.

4. Occupational hazard management at PETROCHEM emphasizes early identification of potential hazards and the integration of efficient means of risk reduction with normal corporate production planning and decision making. Product safety, by comparison, is less well-developed and more reactive to problems experienced by downstream users and customers. In both areas, PETROCHEM argues for external regulation that

40

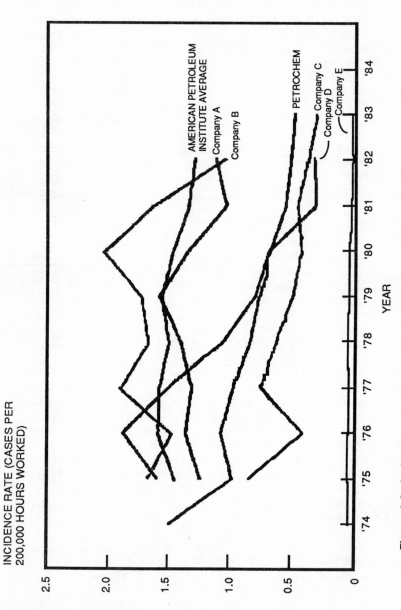

Figure 2.8 Incidence rates — Fatalities plus cases with days away from work.

is performance-oriented and that provides PETROCHEM with the flexibility to choose the most cost-effective responses.

5. By conducting major epidemiologic research on its workers, PETROCHEM assumes the short-term risks associated with the identification of new hazards and the compilation of new health-effects information in exchange for increased ability to manage occupational hazards and to protect the corporation from future liability and burdensome risk-reduction regulations. This is consistent with PETROCHEM's long-term approach to hazard management, its avowed corporate goal to be a leader in industrial health and safety, and the "leverage" of the health-and-safety function within top corporate decision making.

6. Between the two hazard areas—product safety vs. occupational health and safety—reviewed, the latter (with the possible exception of pesticides) is the more extensively developed at PETROCHEM. Present plans to develop a better understanding of product hazards connected with end-use assessment and the institution of a more formal audit system for products should narrow this gap. An omission at PETROCHEM is the lack of a formal product stewardship program, such as exists at a number of other major corporations.

3

Managing Occupational Hazards at a PHARMACHEM Corporation Plant

PHARMACHEM is an international corporation that manufactures, among other products, pharmaceuticals for people and animals. Annual sales exceed several billion dollars, largely derived from the sale of pharmaceuticals. The PHARMACHEM Corporation plant examined in this chapter synthesizes and assembles potentially hazardous materials to produce pharmaceutical products, all the while maintaining low levels of bodily injury and closely controlled chemical exposures to its workers. It does so by incorporating health and safety into the production process; by employing a large and diverse array of control measures; by closely monitoring safety incidents, accidents, injuries, and the use of protective devices; and by capital investment made possible by a highly profitable industry. Unlike PETROCHEM (chapter 2), however; it is still primarily attuned to safety rather than health.

The plant is an integrated pharmaceutical manufacturing unit constructed in the 1970s, covering several dozen acres and valued today at nearly $100 million. It employs hundreds of people, 20 percent of whom are classified as professionals with college or graduate degrees. The plant produces half a dozen product lines. It synthesizes annually several hundred tons of active ingredients for these drugs, employing twenty different solvents, most of which are recovered. The synthesized chemicals are blended with inert materials into some 100 different dosage and package combinations of 1.5 billion tablets and capsules per year.

Production of PRODUCT-A

The plant is a multiple-use facility, which makes major products in "campaigns" during which multiple batches of a product move consecutively through the manufacturing pipeline. PRODUCT-A, a major product at the plant, features campaigns that occur several times yearly. Maintenance, extensive cleanup, and material assemblage and testing precede each campaign.

Following a multistage process in the organic-synthesis division, PROD-UCT-A is transferred in bulk to the manufacturing facility for blending with inert materials, encapsulation, and packaging, in many combinations of size and dosage. Extensive raw-materials, in-process, intermediate, and final-product testing—in all, over 300 specified tests (often with multiple samples)—occurs. Also associated with the production of PRODUCT-A are extensive treatments of waste streams, particularly solvent recovery.

Hazards of PRODUCT-A Production

The occupational health and safety hazards of PRODUCT-A fall into two types: those that are common to manufacturing processes in general and to chemical plants in particular, and those that are specific to the synthesis and manufacturing of PRODUCT-A. In the first class are physical hazards—potentially hazardous releases of thermal, kinetic, and sonic energy arising from the presence and use of heat, machinery, and the movement of materials. Indeed, noise is a significant hazard of the manufacturing process, particularly during tableting and encapsulation. The second class, basically chemical in origin, illustrates the potentially hazardous environment of the plant and the wide range of control actions undertaken to reduce the occupational risk. Table 3.1 describes the major potential releases of chemical hazards.

These releases of energy (e.g., fire, explosion) or materials (e.g., vapors, dusts) provide common descriptors for comparing hazards of different technologies. This method of describing very different technologies in similar biophysical terms (see chapter 1) rests on the assumption that all technologies can be characterized in terms of their manipulations of energy and materials. The hazards of a technology consist of those releases of energy and materials that exceed levels that human beings, other living organisms, or the built environment can absorb.

There are, of course, no risk-free chemicals. Thirty different chemicals enter into the production of PRODUCT-A, and all are potentially hazardous if sufficiently large quantities are absorbed, inhaled, or ingested, or if sensitive body parts such as the eyes are exposed.

The synthesis process used thirteen solvents, all potentially flammable, easily vaporized, and toxic when inhaled. At least one is a recognized carcinogen in animals. An extremely explosive gas, used in combination with other flammables, and several skin and eye irritants also feature in the process. Finally, PRODUCT-A itself is a major hazard in that its very success as a medicine means that inhalation of even small quantities could initiate a physiological response with serious consequences.

How does the plant cope with this impressive array of specific chemical hazards as well as the more common hazards of manufacturing?

Control Actions for the Hazards of PRODUCT-A

PHARMACHEM uses a wide variety of measures to control the hazards of manufacturing PRODUCT-A. One way to order such actions is to use the causal chain model described in chapter 1. This way of conceptualizing a hazard links the chain of events that release hazardous energy or materials to their potential consequences. Each of these links or stages in the causal evolution of hazard offers opportunities for interventions to prevent, control, or mitigate the hazard. Table 3.2 arrays some such actions for the chemical hazards of the PHARMA-CHEM plant. Each row relates to a different type of hazardous release and each column a different stage in the causal chain of hazard.

The list of control actions, derived from managerial interviews and our direct observations, is exemplary rather than definitive. The plant is a modern one constructed in the 1970s and thus incorporates many features that make for a safer and healthier workplace. Most of these features have become standard for this type of plant, so plant management no longer identifies as hazard-control actions certain features that contribute to hazard reduction.

In addition to the overall plant construction, a number of actions apply equally to most types of releases and these appear under *multiple releases* in Table 3.2. They include: the relatively new, modern plant; the careful integration of health and safety information and procedures along with the standard operating procedures that serve as the detailed "cookbooks" for drug synthesis and production; the rigorously enforced ban on smoking in most areas of the plant; the universal use of appropriate personal protection; routine, repetitive safety inspections; and the emergency procedures, medical treatment and compensation programs available to all workers. The various types of releases lend themselves to specific control actions, examples of which are given in Table 3.2

Functionally, these control actions seek to eliminate or to substitute for hazardous processes or chemicals; to contain, segregate, or dilute releases; to segregate people from releases; to monitor releases as a prelude to further action; and to treat and compensate for consequences as they arise. Thus, toluene substitutes for the more hazardous benzene, whereas newer and quieter encapsulating machines will reduce noise levels. Continuous containment shields people from contact with solvents and caustic materials; natural ventilation helps dilute the buildup of vapors; and protective clothing (including fully pressurized suits) segregates people from exposure to toxic materials, dusts, and vapors. Especially hazardous processes are segregated to minimize exposure, as in the use of a separate building for step 1 in the synthesis of PRODUCT-A. Emissions are monitored in six of the production processes and for eleven of the thirty chemicals used, and workers receive annual physical and thrice-annual blood tests.

Occupational Hazard Management

Management structure. PHARMACHEM, like PETROCHEM, is strongly goal-oriented. All levels of organization adopt annually a set of objectives for the coming calendar year. Thus, in safety and health, the individual plant also develops annually its own set of objectives and target activities to serve as a guide to the year's activities.

The goals and objectives of the plant are related to PHARMACHEM's basic corporate safety and health policy. The plant studied has four major objectives: (1) to lower the lost-time incidence rate even below the corporate objective, (2)

Table 3.1

POTENTIAL CHEMICAL HAZARD RELEASES
IN THE PRODUCTION OF PRODUCT-A

PROCESSES	RELEASES
Organic Synthesis	
1. A base material (with a high explosivity rating) in solution with a flammable solvent is reduced to Intermediate 1 by the use of a highly explosive gas and a metalic powder which, in the dry state, can ignite.	Fire, Explosion, Toxic Vapors
2. Intermediate 1 is treated with a caustic liquid alkali and a caustic acid to create Intermediate 2, which is then crystallized, filtered, washed, and dried.	Caustic Materials
3. Intermediate 2 is reacted with a toxic material in an aqueous acid and then treated with a caustic alkali yielding Intermediate 3, which is crystallized, milled, and dried.	Caustic and Toxic Materials
4. Intermediate 3 is reacted with a respiratoxic acid (that reacts with water) and a poisonous (hemolytic agent) liquid and then treated with two flammable solvents producing Intermediate 4, a skin irritant.	Fire, Toxic Materials and Vapors
5. Intermediate 4, a skin irritant, is treated with a flammable respiratoxic gas and a flammable solvent under pressure, filtered, and dried; then purified with an irritating, potentially hepatotoxic and embryotoxic liquid; then filtered, crystallized with a solvent, granulated, centrifuged, and dried.	Fire, Toxic Materials and Vapors

Process Step	Hazards
6. Intermediate 5 is synthesized from a new base material when treated by a solvent, a corrosive acid, and a liquid that is an eye irritant, also yielding a byproduct. The reaction mixture is washed with a solvent carcinogenic in animals), filtered, purified, and dried. It is then redissolved in another solvent and crystallized with the aid of still another solvent.	Fire, Caustic and Toxic Materials and Vapors
7. PRODUCT-A is synthesized from various intermediates with the aid of solvents and is a potential hazard since inhalation of even small quantities can institute adverse physiological consequences.	Fire, Toxic Materials and Vapors
Solvent Recovery	
1. The production of PRODUCT-A involves 13 different solvents, drawn from an underground tank farm containing 57 tanks of 5,000-10,000 gallon capacity. For both economic and environmental considerations, a major attempt is made to recover solvents for reuse by various distillation methods. Some solvents are hazardous and flammable.	Fire, Explosion, Toxic Materials and Vapors
Manufacturing	
1. Sucrose and cornstarch are blended and milled with PRODUCT-A, wet granulated and dried. The mixture is then milled to desired particle size and mixed with a lubricating agent.	Toxic Dusts
2. After sampling and testing, PRODUCT-A is encapsulated in preprinted capsules.	Toxic Dusts
3. After encapsulation, PRODUCT-A is further sampled, tested, and then packaged in varying size and dose combinations.	None Indicated

Table 3.2

CONTROL ACTIONS BY HAZARD RELEASE AND CAUSAL STAGE

Type of Hazardous Release	Change Technology	Reduce or Prevent Release	Reduce or Prevent Exposure	Reduce or Prevent 1st Order Consequences	Reduce or Prevent Higher Order Consequences	Mitigate or Compensate Consequences
Multiple Releases 1-7 S MP	Modern plant with built-in closed circulation	------Standard operating procedures------ Smoking bans	include safety instructions, Personal protection	First aid, Evacuation, Annual physicals, 3 per-annum blood pressure tests	Medical treatment	Reduced work assignment, Workmens' compensation, Life insurance
Explosion 1-5 S		Separate building with high pressure and solvent detectors, knockouts, and alarms. Minimize time (3 days) that explosive gas is on site	Separate building with explosion panels. Evacuation procedures	First aid	Medical treatment	Reduced work assignment, Workmens' compensation, Life insurance
Fire and Heat 1, 4, 5, 6, 7 S		------Fire fighting training and equipment------ Special materials packaging, handling and loading	Evacuation procedures	First aid, Evacuation	Medical treatment	Reduced work assignment, Workmens' compensation, Life insurance

49

Hazard	Substitute	Solvent detectors	Open building	First aid	Medical treatment	Reduced work assignment
Toxic Vapors 1, 4, 5, 6, 7 S	Substitute toluene for benzene, reduce chloroform	Solvent detectors 1 Emissions monitoring 1, 4, 5, 6, 7, S Closed solvent circulation 1, 4, 5, 6, 7, S Rooftop charging 4	Open building with natural ventilation	First aid Evacuation 3 per-annum blood pressure tests	Medical treatment	Reduced work assignment Workmens' compensation Life insurance
Toxic Materials Dusts 3, 4, 5, 6, 7 S MP		Isolated blending facilities MP	Fully pressurized protective suits 4, 5, 7, MP blood pressure tests, worker rotation	First aid Evacuation 3 per-annum blood pressure tests	Medical treatment	Reduced work assignment Workmens' compensation Life insurance
Caustic 2, 3		Materials in closed tanks 2, 3, 6. Special handling, charging 6	Acid protective suits 6	First aid Evacuation	Medical treatment	Reduced work assignment Workmens' compensation Life insurance
Noise	New encapsulating machines MP	Machine enclosures MP	Personal protection MP Worker reassignment MP	Hearing Conservancy Program Annual physicals	Hearing Conservancy Program Medical treatment	Reduced work assignment Workmens' compensation

1-7 - Synthesis Stages   S - Solvent Recovery   MP - Manufacturing Process to assure continued compliance with OSHA noise and hearing-conservation regulations, other OSHA regulations, and the hazard communication rule, (3) to implement various other administrative actions related to fire prevention, safety, and chemicals in the workplace, and (4) to realize savings in safety expenditures.

The first goal relates to the common standard used to measure progress in industrial safety—the incidence rate for events with the potential for worker injury and the OSHA-reportable, lost-time incidence rate, composed of actual lost-time accidents and occupational illnesses. PHARMACHEM adopted as an objective a rate of 1.94 and the plant sought to better that, setting for itself an internal goal of 1.48. Achievement of this goal in the year of study represented a three-fold improvement in lowering the incidence rate since plant start-up in the early 1970s.

Other major objectives center on assuring continued compliance with regulatory agency requirements, particularly noise and hearing-conservation regulations and the hazard communication rule. Continued plant participation in the corporate "Chemicals in the Workplace" program also figured prominently.

In terms of management organization, the health and safety of the employees is the direct responsibility of all managers in the plant, and four full-time employees (two nurses, two safety officers) have specialized health and safety responsibilities. Occupational health, it should be noted, falls under the jurisdiction of the personnel manager, a practice that has been phased out at many corporations and that reflects an older view of health service as part of labor relations rather than of hazard management. The plant employs two full-time industrial nurses and a part-time physician who visits the plant on a regular basis and has training in occupational health. A consultant has been employed at intervals to review the annual physicals and incidence reports for trends suggestive of hazardous situations. The consultant's report is key to the plant manager's own periodic review of such records, which also seeks to identify potential hazard trends. The PHARMACHEM plant, unlike the individual PETROCHEM plant, discussed in chapter 2, employs no industrial hygienist.

Occupational safety currently resides within the Technical Services Department and consists of the chief safety officer, a mechanical engineer with considerable experience (including reactor safety with the Atomic Energy Commission), and a safety officer trained as a fireman. Typical activities of the two safety officers include training, daily and monthly inspections and their follow-up, investigations of incidents, accidents, and potentially hazardous situations, and monitoring, record-keeping, and report-writing.

Insofar as all managerial personnel ostensibly share responsibilities for safety and health, management meetings include discussions of health and safety matters. In addition, the plant has departmental safety committees of managers and workers and a plant-wide safety committee that conducts the monthly plant inspection. The plant manager has direct responsibility for safety; he receives immediate reports of accidents and examines in person the daily incidence records. From a detailed review of his commitments, he estimates that he devotes about five days per month, or 20 percent of his working time, to overall hazard management, which includes environmental hazards as well as occupational health and safety.

Technical knowledge about chemicals and production practices is considerable at the PHARMACHEM plant. The industry as a whole is strongly "procedure-oriented," and the demands of careful quality control tend to merge with the behaviors necessary to prevent either releases of toxic materials or careless operations. A staff that is capable of conducting hundreds of repetitive quality-assurance tests can be expected, when instructed to do so, to monitor effectively chemical releases of various types.

Prior to the production of PRODUCT-A at the plant, the standard operating procedures developed in pilot runs in the laboratory and other plants included detailed safety and health cautions. Material safety data sheets (MSDSs) were provided using PHARMACHEM own data sheets or standard industry or Chemical Manufacturers Association data The potential for explosions of intermediate chemicals in some of the reactions underwent separate testing. Finally, PRODUCT-A itself was subjected to intensive study, using the data provided in the drug-efficacy and safety-approval process.

To assist local facilities in managing occupational hazards, the corporation produces a *Facility manager's guide to safety and health*, a compendium of goals and objectives, program formats, training programs, and record-keeping procedures for organizing safety and health functions, and *Safety Highlights*, which reports monthly performance records of facilities. Other materials include a monthly report describing how locations are progressing in the "Chemicals in the Workplace" program, now in place in all of the operating plants.

The "Chemicals in the Workplace" program includes the use of area monitoring for many chemicals that have time-weighted average exposure limits published by organizations such as the American Conference of Governmental and Industrial Hygienists (ACGIH) and the National Institute of Occupational Safety and Health (NIOSH), the OSHA code, or both. Both the corporation and the operating plants maintain an up-to-date master chemical list which includes liquids, vapors and gases, and solids (dusts). Testing provides a check of work areas during the use of chemicals in production, cleaning, and maintenance activities. These operations receive checks periodically and also when significant process/operational changes occur. New chemicals used in the plant are added to the list and are subject to the testing program.

In general, the role of the corporation, as compared with that of the plant management, entails setting goals and supplying information. We were not able to identify the conditions under which corporate headquarters would intervene in local plants, nor was it clear whether or not a regular schedule of health, safety, or environmental audits exists (in sharp contrast to the other corporations studied in this volume).

Management functions. The management functions in hazard management entail a five-step process operating on the causal chain of hazard as shown in the flow-chart (Figure 1.2) in chapter 1. Management seeks to alter the flow of events in the causal chain by blocking or reducing hazards at each of the stages through appropriate control actions. The first step—*hazard assessment*—identifies hazards, measures their risk, and judges their tolerability. For risks that are not tolerable, *control actions* are sought and, in the choice of these, a strategy is *selected, implemented,* and then *evaluated.*

In PHARMACHEM as a whole and at the plant studied, these managerial functions merge. To the pragmatic hazard managers of PHARMACHEM, control analysis is an intimate part of hazard assessment and management. Discussion of controls, and their analysis and selection, is embedded in hazard assessment documents. Management strategy appears to emerge both from overall corporate policy and from the particular professional background of the plant managers.

Prior to the manufacture of PRODUCT-A at the plant, laboratory production, pilot bulk lots, and actual production campaigns were undertaken at other PHARMACHEM plants. Hazard assessment, consisting of identification of safety considerations for the handling of raw materials and intermediate processes and the formulation of recommendations for personnel protection, was performed at these pilot sites. To prepare hazard-management procedures, five raw materials not previously used in the plant initially producing PRODUCT-A were analyzed for their potential hazard to workers; five steps were singled out in the synthesis process for special attention; and the milling and handling of PRODUCT-A itself underwent careful analysis. In-house studies of possible dust explosions accompanied the analysis, as did the use of corporate and industry MSDSs.

These assessment materials were merged into an overall (but incomplete) risk assessment, consisting of descriptions of consequences but lacking estimates of the frequencies or probabilities of hazardous events. The risks are judged tolerable to the corporation on the basis of such reasoning as "established shop safety procedures are available to cope with all the above hazards" and "pilot plant experience...gives no reason...to anticipate any special processing hazards beyond those presented..." In the case of a material with a severe dust-explosion potential rating, the risk is judged tolerable because (1) the minimum ignition energy required is far above the level produced by electrosotatic sparks from grounded equipment, and (2) the changing of the solid is permissible only in areas where dust can be easily controlled. Thus, in both cases, risks are judged tolerable because, in the managers' view, the hazards can be coped with by appropriate and well-understood control actions.

The long experience with the handling of hazardous chemicals and the various pilot operations for designing a manufacturing process provides an initial hazard assessment. But much hazard assessment also derives from extensive experience in actual manufacturing operations and in the particulars of the plant. Thus, an ongoing process of speedy monitoring, screening, and diagnosis is very much a part of hazard assessment. Monitoring takes the form of solvent and high-pressure detectors and a network of air-sampling monitors. Screening, consisting of annual physical exams and thrice-annual blood tests for potentially exposed employees, is employed in the limited health-surveillance program. Diagnosis is based upon the screening results, the review (by an outside consultant) of health data, the careful analysis of all incident occurrences, and repeated plant safety inspections. The plant manager is ultimately responsible for identifying and acting to remedy any hazardous trends.

The analysis of hazard controls in PHARMACHEM and the plant is a sequential but iterative process. As expressed by local management, corporate policy prefers certain controls, favoring the elimination of hazardous conditions or chemicals over engineered safety, but also favoring engineered safety over behavioral adjustments, including personal protection. This ordering of control actions is based on management's assessment of relative effectiveness, and the search for a control package follows this preferential ordering.

Thus, when the carcinogenic hazards of benzene became more widely acknowledged, toluene, found to be less hazardous, was substituted where possible in organic synthesis. Given the widespread need for solvents, the preferred control measure is closed circulation, backed up by the natural ventilation of the open-design structure of the organic synthesis plant, the monitoring of all the major solvent emissions, and the use of personal protection in specified situations. The hazard managers do not comprehensively assess the entire array of possible control actions. Rather, they take what they consider the most conservative tack—they first examine only the preferred alternative to see if it is effective and economically feasible; if not, they move on to other alternatives.

PHARMACHEM is a highly profitable corporation with a very large investment in research and development, so that its manufacturing costs are proportionately lower than those in other industries. Assuring the quality of its product and maintaining the reliability of supply of the product are driving priorities. In such a context of high profitability, strong quality control, and a secure flow of product, the marginal costs required for health and safety protection do not loom large. Some merging of management and workers' interests occurs in the desire to prevent potential accidents or releases of toxic material that not only injure workers but may compromise the quality and flow of the product as well. Thus, consistent with production goals, most marginal investments in safety and health are readily undertaken, although major investments appear to receive the same scrutiny that any profit-conscious corporation would undertake.

Avoiding and reducing risk are the preferred strategies for hazard management at the plant. But both workers and management accept, or at least tolerate, some risk. Hence the need exists for programs that mitigate hazard consequences.

A central limitation, however, exists in the differential handling of occupational health and occupational safety. PHARMACHEM in general and the plant studied in particular are more attuned to reducing the risk of accidents than the risk of disease and impaired health, to attending more to the risk of acute hazard events than to chronic events, and to coping with the immediate rather than the delayed. At the plant, this is apparent in the differential corporate organization of health and safety (health is a personnel service—although this is to change in the near future) and in the frequency of inspections, committees, and statistics of achievement for safety as compared to the more limited activities related to the long-term dangers to health from chemicals.

The implementation of hazard-management control actions begins with the plant's annual set of objectives, each of which contains a target completion date. Plantwide inspections (monthly), near-miss and incident inspections, and monitoring activities provide ongoing routinized management activities and generate special hazard-management initiatives.

The bulk of day-to-day activity consists of housekeeping and training, dominated not by the special characteristics of the hazardous chemicals being handled, but by the common characteristics of people and production work everywhere: injuries to hands and feet, falls, slips, and strains. Safety inspections focus on the state of emergency preparedness: fire extinguishers, signs, escape routes, electrical grounds, and the like. A few points come in for some special emphasis: corrosion and sparks are constant enemies in a plant where chemicals need to be contained and explosions prevented.

Even in the best-trained and most attentive work force, noncompliance with health and safety requirements, particularly in the use of personal protective devices, is a recurrent problem. Although the apparent widespread use of personal protection at the plant is impressive, observations of noncompliance were also absent in the plant's safety documentation. Management informed us that when a worker violates a safety or health practice, the immediate supervisor is held responsible for obtaining compliance. Such cases are reviewed at safety meetings and dealt with by an oral report or by other supervision, as well as, it is claimed, by peer pressure among workers.

Two types of evaluation are evident in the procedures at the plant. The first includes *outputs,* consisting of the targets that have been met, the meetings held, the number of workers participating in training sessions, and the like. The second relates to *effects,* particularly the widespread use of the OSHA lost-time accident and injury rate. This single number, despite its evident limitation, functions as the "bottom line" equivalent for occupational health and safety, much as the production figure, sales, or profitability numbers serve for major corporate economic goals.

Drills, particularly surprise drills (held only infrequently), serve as training for emergency procedures. Alternatively, near-misses serve as natural experiments, and careful investigation of such incidents provides a continuing form of evaluation. In the areas of health and safety, however, no independent evaluation processes equivalent to the quasi-independence of the quality-assurance department in product-reliability control appear to exist, nor, as we have already noted, is a regular audit policy apparent.

Conclusions

The PHARMACHEM plant studied in this chapter combines the general occupational hazards of a small-parts-assembly plant and warehouse with the special hazards of chemical storage, synthesis, and blending, thereby posing occupational risks of toxic materials, fire, and explosion. PRODUCT-A, the case-study product, is itself potentially hazardous. As a powerful drug in very small quantities, it is a significant potential threat to workers who manufacture it in large quantities. Involved in its production are 30 chemicals of varying degrees of hazard, all posing some threat through accidental release, combustion, absorption, inhalation, or ingestion.

Despite the special characteristics of these chemical hazards, recorded incidents at the case-study plant suggest a pattern of injury and illness related much more to common general occupational hazards than to the special qualities of chemical use and pharmaceutical manufacture. This is due in part to the long existing bias of corporate hazard managers towards safety rather than health, acute rather than chronic conditions, and immediate rather than delayed effects.

The structure, organization, resources, and control actions of the case-study plant all reflect this bias. Nevertheless, the plant does display a diversity of hazard-management approaches and control activities directed at chemical hazards. These control actions surely contribute to the overall low injury-incidence rate at the plant and the small number of chemically related incidents. These actions, which take place at all stages of the causal chain of hazard, use changes of technology, and measures to prevent releases, to minimize exposure, to prevent consequences, and to mitigate them after they occur. As in most hazardous industries, the greatest effort is allocated to preventing releases and potential exposure to them.

PHARMACHEM and the plant studied appear to prevent or eliminate risk where possible at moderate cost and to respond promptly to regulatory requirements. Its managers tend to rely, in the main, on outside information about hazard, seeking internal solutions to specific production problems. They are content to be in the middle ground of performance compared to other industry leaders (*cf.* with PETROCHEM and Volvo, chapters 2 and 4).

In industry as a whole, the standards of success in hazard management are not well developed. The one widely used industrial standard, the OSHA reportable-incidence rate, is also used at PHARMACHEM and the case study plant. By that standard the chemical industry's rate is well below that of industry as a whole. PHARMACHEM is close to the mean for the pharmaceuticals industry and the case study plant is better than that, having reduced its lost-time incidence rate threefold in a decade.

How successful the plant is in protecting the long-term health of its work force is uncertain because the health-surveillance program is rather rudimentary and has functioned only over the limited life of the plant (since the 1970s). Long-term occupational diseases, if any, have not had time to show themselves.

Still another measure of managerial success is compliance with regulatory procedures. In general, anticipatory compliance and many safety and health measures meet or exceed regulatory requirements.

Contributing to the hazard management programs is the high profitability of the industry in general and of PHARMACHEM in particular. A highly profitable enterprise can, and often does, readily absorb the often small additional investments required in the health and safety of the workplace. But, in addition to its ability to undertake investment in health and safety, the pharmaceutical industry appears to enjoy a unique convergence between hazard-management objectives and the production and long-term profit goals of the industry. The prescription-drug industry promises to its customers—pharmacies, doctors, and their patients—a high and dependable quality and supply of what are potentially dangerous, albeit beneficial, substances. This is an industry where the beneficial and harmful qualities of chemicals are always being manipulated, where high standards of quality product are demanded, and where the security of product availability is crucial. A high ratio of professionals to nonprofessionals suggests a relatively high knowledge of chemical behavior. All these features readily blend with, and support, hazard-management programs.

Finally, distinctive local features characterize hazard management at the plant. It is, after all, a new plant incorporating many desirable features of closed circulation, ventilation, containment, and isolation. To the visiting hazard researcher, the plant impresses with its potential hazardousness and its actual achieved safety. Comparable assurances for the long-term health of its workers are less certain and await the test of time.

4

Managing Product Hazards at Volvo Car Corporation

Ola Svenson
Lund University

The first Volvo automobile made its debut in 1927 and through the early 1930s the company made only a couple of hundred cars each year. Since 1928 when the first Volvos were exported, the export market has become the most important market for the company that is Sweden's leading exporter. Today, Volvo is an industrial group manufacturing many products, which represent about one-tenth of Sweden's total exports. The Volvo Car Corporation has been profitable through the years.

In addition to leading the main exporters of Sweden, Volvo is one of the most export-oriented transportation industries in the world, with as much as 75 percent of its total sales in export markets. From an international perspective, Volvo is small in terms of numbers of cars produced, but it is a much larger manufacturer of trucks and busses in percent of the world market. More than 100,000 shareholders own Volvo, and the ten largest together command approximately one-fifth of the votes. The largest individual owner, the Fourth National Pension Insurance Fund, owns some 6 percent of the shares. Yet, economic interests closely affiliated with a Swedish bank (Handelsbanken) together command a somewhat greater proportion of the shares.

The companies belonging to the Volvo group produce annually about 350,000 cars in Sweden, Canada, Belgium, and the Netherlands, of which some 180,000 are built in Sweden. The cars produced in the Netherlands belong to the smaller car series whereas those produced elsewhere are cars of the 760 series and 1300-kg cars of the 240/260 series that will be the focus of this chapter. About 75 percent of the parts of that car are manufactured by independent subcontractors but all the 240/260 cars are produced on Volvo's own assembly lines in Sweden or abroad.

Volvo is particularly noteworthy as a leader in automobile safety. Over the years, it has generated and marketed many of the safety innovations that have ultimately found their way into regulations in Sweden and abroad. Of the various corporations treated in this volume, Volvo has been a safety leader over the longest period of time. The analysis that follows examines Volvo's hazard-management structure and process within a broad societal context.

Hazard-Management Organization

The organizational structure of the Volvo Car Corporation, with an emphasis on hazard management, is shown schematically in Figure 4.1. As the diagram makes clear, many parts of the organization handle quality and safety aspects of the cars. The coordinating unit for safety and environment belongs, interestingly, to the Department of Quality. The crashworthiness and crash-avoidance investigations are performed in the Volvo Safety Center, which belongs to the Department of Product Development and Design. No specific organizational body is devoted exclusively to hazard management, unlike the case of PETROCHEM (chapter 2).

Safety and Environment Department. Generally speaking, this unit is responsible for developing Volvo policy and organization to conform to legal requirements (e.g., engineering requirements and product-liability laws) regarding safety of the cars produced. The unit represents the company in deliberations with national agencies concerning the safety questions of cars of the 240/260 series. It also collects legal and type-approval requirements in different parts of the world and analyzes and distributes the information to all relevant units in Volvo. Furthermore, the Department is responsible for compiling technical specifications needed in product-liability claims (see below) and for Volvo's presentation of that evidence in case of a court trial. The Department also represents the company when arranging for type approval and type certification of the cars.

A very important instrument, issued by the Safety and Environment Department, is the *Legal requirements design manual* that provides all the necessary information about existing laws and technological specifications from the whole world. In addition, its basic files contain results, applicable to such legislation, of safety and environmental research from different countries. The Department continuously updates the manual and distributes information throughout the company for guidance when, for example, changes occur in the design or specifications of cars.

Within Volvo, the Safety and Environment Department is responsible for establishing and maintaining the required posture for product liability and for coordinating the overall product-liability prevention and reduction program. The preventive activities consist of: (1) total quality assurance, including vendor quality control, (2) design review programs, (3) special marketing and handling programs for safety and emissions systems and components, (4) legal-requirements tests, and (5) individual reporting responsibility for product defects (delegated to the lowest echelon). Restorative activities include rapid retrieval of documentation pertaining to different units within the Volvo organization, service network corrections and fast repair when required, and competent engineering representation in legal proceedings.

Environmental protection, involving emissions and fuel economy, noise, and scrapping with material reuse, is a comparatively new area for the automobile industry. Exhaust emissions and fuel economy are particularly important today, but in contrast to tests of crashworthiness and crash avoidance, car producers do not have a long research tradition in this field. Therefore, some of the competence and information at this stage must be obtained from outside the company. It is necessary to contend both with regulated pollutants (e.g., HC, CO, NO_x) and nonregulated pollutants where the hazardous substances are largely unknown. Emissions and fuel economy are commanding increasing interest within the company.

The Safety and Environment Department comprises about 15 persons whose work is organized into different sectors concerned with (1) legal requirements, type approval, and type certification, (2) fuel economy, (3) product liability, (4) documentation, and (5) coordination with Volvo Car BV (Holland). The goal is to keep the Department small so that an ease of communication in a small group will prevail. Since the early 1970s, the Department has grown gradually, with clear support from the Volvo management. Typically, the people working in the department are younger middle-aged persons who tend to stay in the Department, thereby providing continuity. Most are technicians, although one has a degree in law. The people working in the Department seem well informed, have support from the company, and work with many informal contacts.

The Volvo Safety Center. Whereas the Safety and Environment Department mainly analyzes information in documents, reports, and journals, the Volvo Safety Center also generates a great deal of research data bearing upon hazard identification and assessment. The work of the center includes crashworthiness tests and accident investigations. The latter provide a remarkable system of "feedback" on the product subsequent to its dissemination. About 30 people who work in this unit belong to one of four groups: (a) management coordination and analysis, (b) experimental work, (c) accident investigation and (d) maintenance and work shop. The unit plans all the crash tests performed by Volvo and works in close cooperation with the designers (who are in the same building). The technical facilities include a crash track, a smaller cabin accelerator, and facilities for ergonomic testing.

The chief aim of the accident-investigation group, which has been in place since 1965, is to gather facts of importance to the product development of new cars. In particular, the investigations address questions about why drivers and passengers are injured in accidents, physical limits of human tolerance, accident sequence, and consequences of accidents in relation to vehicle and driver performance.

Since 1970, the group has conducted thorough investigations of all accidents within one hour's travel by car from the Safety Center, thereby covering the city of Göteborg and its surroundings. The investigation of an accident starts as soon as possible after the police have been notified and is performed by two members of the group of four engineers. The investigations, in effect, are case studies on the spot. In addition, a medical doctor is part of the group but does not participate in the on-the-spot work. The medical part of the investigation of an

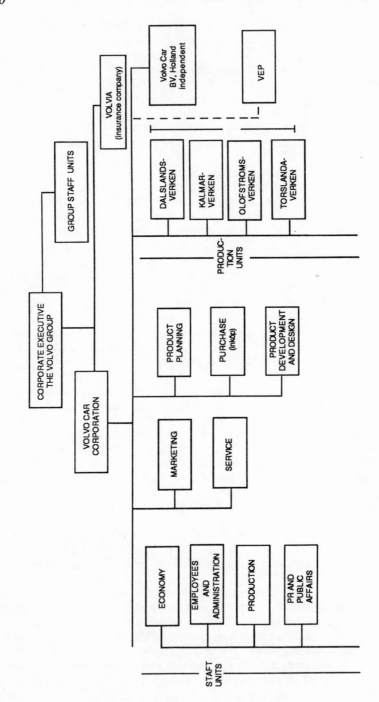

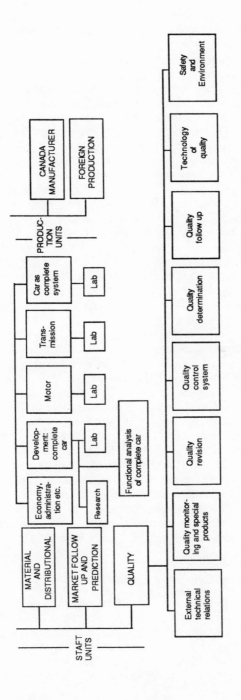

Figure 4.1 Organizational structure of Volvo, with emphasis on hazard management.

accident starts no later than seven days after the accident and includes monitoring, for six months following the day of the accident, the condition of the people injured.

When investigating an accident, the personnel of the accident-investigation group use the tools commonly employed in such on-the-scene investigations. They also use a standard questionnaire to determine the causes leading to the accident and the physical and human particulars contributing to injuries of the occupants. Data on about 300 different items are gathered from each accident, stored in chronological order, and classified by key-words. This renders it possible to sample all accidents characterized in a certain way in response to particular questions raised at any time. Although well over 1000 accident investigations have been conducted, the author knows of no overview of the results and findings from all these cases. Nevertheless, reports covering parts of this rich data base are published continuously. Also, it is possible to estimate annual costs of the accident investigations of at least 1.5 million Swedish Crowns (US $175,000). Additionally, the Volvo insurance company has performed numerous statistical analyses.

The Recall Committee. Founded in 1972 by the Department of Quality, the Recall Committee comprises people from different parts of the Volvo organization—specialists in quality, design, production, service, spare parts, and law. Since 25 December 1974, the car manufacturer must pay for the repair (recall) of defective vehicles less than eight years old.

Over time, the Committee has handled annually about 200 deficiency reports, of which about 10 qualify as potentially serious and deserving further investigation. In recent years, however, the number of potential cases has dropped considerably (to less than 100 reports a year). Still, about 20 reports per year are judged as serious. The reduction of the number of potential cases may depend in part on the awareness in the Volvo company of the Recall Committee.

The information leading to a recall comes from several different sources. The service organization may report a suspected defect to the committee. Two or three recalls occurred each year up to 1979, but between 1979 and 1983 the company made no recalls. In all, a total of about 20 recalls have been made over the years. Volvo made the recalls on its own initiative; legal directives were never required. A Volvo recall in any market includes other markets having cars with the same defect thereby assuring uniformity at all company locations (an issue pursued in the Union Carbide case in chapter 6). The types of recalls involved range from quite small details judged unimportant by most owners to more important ones.

Safety of Volvo Cars and Company Behavior

General Volvo policy. As mentioned earlier, Volvo is a very large company and quite important to the Swedish economy. Correspondingly, Volvo is also quite dependent on the rules for Swedish trade and economic activity. This may explain why the company actively participates in the current economic debate and other issues of national interest. Volvo also presents its official view on the role that the automobile plays in society. Traditionally, safety issues have been viewed as important, as the following official statement by Pehr G. Gyllenhammar, the president and general manager of the Volvo Group of companies, makes clear:

- Volvo does not intend to protect motor vehicles at any price and in all connections.
- However, the motor vehicle is indispensable today as a transport unit.
- It is in the interest of Volvo that motor vehicles are used in such a way that they do not cause damage or injury.
- Volvo now considers its responsibility to be not only to ensure that the products are practical as transport units but also that they function in the widest perspective—in our environment.
- Volvo alone cannot solve the environmental problems associated with motor vehicles. The society carries the main responsibility for developing our transport systems. But Volvo is determined to make active contributions with viewpoints and proposed solutions.
- Volvo is convinced that a city environment which is both vital and favorable to human beings can be combined with efficient transport resources. The society needs both.
- Volvo considers that neither fantastic Utopian products nor a romantic back-to-nature movement will solve the problems of the society but believes instead in practical and simple solutions which can be discussed and understood by everyone.

Volvo and safety in the past. When Volvo manufactured cars in the 1930s and 1940s, the roads were poor and the climate was (and still is) harsh in Sweden. These realities made it necessary to build strong, robust cars, and quality became an important aspect of automobile manufacturing. People at the time may have been more quality minded than today. Volvo initiated early important quality requirements for automobile safety, including laminated windshields, introduced in the 1930s, and (later) windshield defrosters. Of course, other cars had these features as well but the details were not regulated by law anywhere in the world at the time. Central to hazard management at Volvo, as at PHARMACHEM (chapter 3), is the close relationship between quality and safety: safety is conceived as one of the key aspects of quality. In the 1940s, safety became an important characteristic in itself and was considered to be an effective appeal in marketing Volvo cars (a judgment generally not shared by American automobile manufacturers, although recent advertisements have begun to play up safety.)

The safety profile of the company became more pronounced during the late 1950s when Volvo introduced seat belts and padded dashboards in all its cars. Why safety became such an important feature for the people of the Volvo company is uncertain. Perhaps the Scandinavian mentality coupled with a general manager (Engellau) who—much as PETROCHEM's Vice President for Health, Safety, and Environment (chapter 2)—personally was convinced about the importance of safety? Perhaps Mrs. Engellau, who was a physical therapist, could inform her husband and other Volvo people about the consequences of an accident in a very graphic way? These and other speculations come to mind when studying a company that sells cars with a whole section of their sales brochures illustrating what happens in a collision. Most manufacturers have tended to skip that embarrassing possibility, while stressing all the nice things about their product. Of course, the outcome of an accident in a Volvo may be as favorable as can be, but that does not explain the whole story. It suggests something about Swedish mentality and about the quality considerations that are important for Swedes who buy Volvo cars.

Table 4.1 compares a sample of safety requirements introduced in Volvo cars with legal requirements introduced in Sweden and the United States. As the table shows, Volvo has consistently been a little ahead (e.g., head restraints, seat anchorage) or had the safety features (e.g., padded dashboards, seat belts) very long before any legal requirement. Whereas other well-managed corporate safety programs—e.g., at PETROCHEM and Rocky Flats (chapters 2 and 5)—initiate some safety requirements in advance of regulation, Volvo is clearly the most striking example over the longest time of such an industry leader.

Hazard-management philosophy at Volvo. Volvo uses a systems approach to hazard management. This approach was explicitly formulated by the head of the Safety and Environment Department, Lauritz Solberg Larsen, in 1975 in a paragraph entitled "The Systems Approach: Cause and Effect":

> Today we consider normally three causative elements in the road safety system, the driver, the vehicle and the road itself, with its environment such as signs and signals. More generally, we speak of a man-machine-milieu system and can treat each element's involvement in transportation in a rational way. (Larsen 1975, 42-3)

This way of looking at accidents closely resembles the way of classifying a crash into three stages for three subsystems, as shown in Figure 4.2 (*cf.* Haddon 1972) as well as with the "hazard chain model" used in this volume (see chapter 1). For identification and correction of road hazards, the systems approach adopted by the Committee on the Challenges of Modern Society gives another general frame of reference for the Volvo hazard-management approach (Figure 4.3).

In 1966 the United States enacted companion legislation in the Highway Safety Act and the National Traffic and Motor Vehicle Safety Act. Volvo essentially adopted both acts, which relied on standards and reflected

> a typical example of the systems approach in which the machine or vehicle is considered as one system having 46 standards (a standard may concern, e.g., hydraulic brake systems). The next division has two subsystems: *crash avoidance* and *crashworthiness*. Under the former have been issued 26, and under the latter 20 standards. Furthermore, additional subdivision has been performed. Under *crash avoidance* we have Vehicle Handling and Stability, 10 standards; Visibility, 6 standards; and Driver Environment, 10 standards. Under *crashworthiness* there are Occupant Compartment, 11 standards, and Vehicle Crash Energy Management, 9 standards. (Larsen 1975, 45)

Table 4.1

SOME U.S. AND SWEDISH LEGAL REQUIREMENTS COMPARED WITH VOLVO STANDARDS AS OF 1981

COMPONENT/SYSTEM	U.S. LEGAL REQUIREMENT SINCE	SWEDISH LEGAL REQUIREMENT SINCE	VOLVO STANDARD SINCE
Split Brakes Systems	January 1, 1968	Model Year 1972	Model Year 1966
Windshield Defrosting/Defogging	January 1, 1968	March 27, 1968	August, 1954
Windshield Wiping/Washing	January 1, 1968	March 27, 1968	August, 1960
Safety Glass	January 1, 1968*	Model Year 1969	1930s
Padded Dashboards	January 1, 1968	Model Year 1970	August, 1956
Seat Belts	January 1, 1968	Model Year 1969 January, 1963 (USA)	August, 1958 (Sweden)
Seat Anchorages	January 1, 1968	Model Year 1970	August, 1967
Fuel System Integrity Step 1 (Frontal Impact Only)	January 1, 1968	Model Year 1969	Model Year 1968
Step 2 (Frontal, Rear, and Side Impact)	September 1, 1976	Not required	September, 1976 (USA)
Windshield Mounting Step 1 (Light-Load Impact)	January 1, 1970	Not required	August, 1969
Step 2 (Heavy-Load Impact)	September 1, 1978	Not required	September, 1978 (USA)
Head Restraints	January 1, 1969	Not required	Model Year 1968

* Some state regulations enacted earlier.

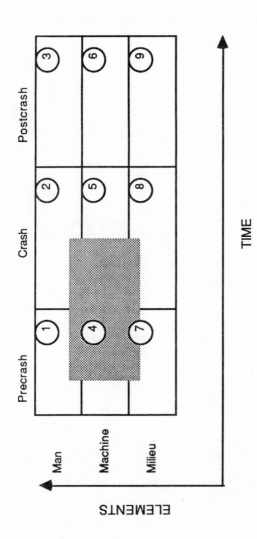

Figure 4.2. Investigative matrix for automobile and accident research in road safety with traffic elements compared with time. The shadowed area represents the concentration of the work on Volvo's Experimental Safety Vehicle.

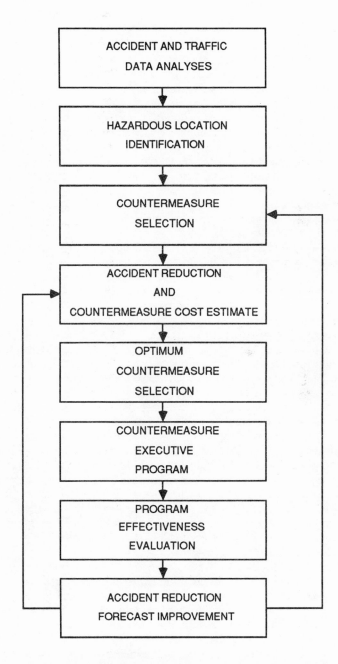

Figure 4.3   Volvo's feedback system for identifying and correcting road hazards.

Crash-avoidance engineering. The company exhibited the Volvo Experimental Safety Car (VESC) during the Third International Conference on Experimental Safety Vehicles in Washington in 1972. Since then, Volvo has developed the VESC continually and incorporated safety components and systems derived from this work into the cars in production. The existence of a safety target was very important for the engineers at Volvo, for those working with the VESC, and for those improving the standard cars. One of the requirements that the VESC should meet was that its handling characteristics should change as little as possible, especially in different emergency situations. Thus, the car should permit a large percentage of the drivers to form a man-machine system that is controllable in critical situations (*cf.* Jaksch, Gustafsson, and Solberg Larsen 1974). Measuring the only partly perceived handling characteristics of a driver-vehicle system has been a great difficulty for car inspection agencies and auto manufacturers alike. The report by Jaksch *et al.* (1974) presented suggestions and empirical data on ways to measure vehicle characteristics in terms of maneuverability (Jaksch 1979a, b).

Crashworthiness engineering. Designing a car with a high degree of crashworthiness means that it should have an occupant compartment that withstands crash forces and prevents injury of the occupants. Furthermore, the car and the compartment should absorb or manage vehicle crash energy so that the occupants are subjected to acceleration-time exposures that do not exceed human tolerance limits. Lastly, the car should cause as little damage as possible to other vehicles and traffic participants should an accident occur. Volvo has spent substantial effort in designing crashworthy cars, as indicated by its progress on deformation zones and seat belts described below. Each year about 70 full-scale tests, corresponding roughly to 40 tests per 100,000 cars produced, are made in the Volvo Safety Center crash track. Several hundreds of tests are also made on the cabin accelerator where the designs of specific safety details are tested.

With other researchers in the field, Volvo has found strong arguments that point to a low-force level, long-deformability, energy-absorbing front part of the vehicle instead of a high-force level, short-deformability system to absorb the kinetic energy in crashes. The design of the 240 and 260 series of cars illustrates this standpoint. Informally, Volvo estimates one of the safety benefits in introducing the 240 series with a long deformation zone to be about seven lives saved per 100,000 cars in use. As mentioned above, the existence of safety prototypes appears very important for reducing hazards. Other industries might well profit from introducing the safety prototype idea (such as the inherently safe reactor proposed for the nuclear power industry).

During the 1950s, engineers at the Swedish State Power Board started testing seat belts by having cars dropped to the ground with belted dummies. The purpose was to increase the safety of the drivers of the cars used by the Board. The seat-belt design was inspired by airplane seat belts. In 1958 a Swedish delegate to a meeting of the Economic Commission for Europe (ECE) meeting raised the seat-belt question, but the issue sparked little interest. A year later, however, the atmosphere had changed, and the concern over safety belt questions was quite striking. It may also have been important that the wife of one of the leading personalities of that meeting had had an automobile accident during the year between the two meetings.

In 1959, Volvo began to install three-point seat belts as standard equipment in all their models. Following this introduction, Volvo performed a large-scale study of the effects of seat belts on driver and passenger safety. The results of this follow-up study clearly demonstrated increased safety for drivers and passengers using their seat belts in comparison with those who did not (Bohlin 1967; Volvo 1967). In fact, the results were available to the U.S. National Highway Safety Bureau (NHSB) in advance of the planned date of publication of the report. The information gathered by Volvo was quite important for the NHSB's report to the U.S. Federal Highway Administration—then seeking information about the effects of seat belts—as well as for other informed groups in the United States. At the time, the net positive effects of safety belts were questioned by different parties, including auto manufacturers and the Federal Highway Administrator. Volvo's seat-belt research, design, and experience were also extensively reviewed before federal standards were issued in Australia (Larsen 1975). Bohlin (1977) has also studied the effects of introducing the Swedish law requiring front seat passengers and drivers to wear seat belts when travelling. The issue continues to spark debate in the United States during the 1980s as individual states enact, repeal, or re-enact restraint laws.

Hazard Management and Quality Control at Volvo

Large corporations react to societal regulation by creating routines to cope with societal demands. One of these routines—recall—has already been discussed, and another—product-liability claims—will be treated below. Here a third—the quality control process—is related to Volvo's management of automobile safety.

Volvo defines the quality concept very broadly. The definition includes aspects such as environmental factors, operating characteristics in different situations, reliability of functions of the car, maintainability, fuel economy, etc. Some of the quality aspects are directly safety-related and are very important for hazard management. Naturally, quality control is applied at all stages within the car manufacturing and maintenance process. Roughly 10 percent of the total time spent by people on the assembly line manufacturing a car is devoted to quality control. General quality control is performed in order to fulfill specified requirements that may originate from customer, product specifications, safety requirements, etc. Given limited resources, the priority order between attending to different types of quality aspects was established in 1978:

> In setting priority because of limited resources, the resources available shall first be applied to solve safety and legal requirement problems, and next to solve customer irritation problems. The measures aimed at solving problems which initiate high warranty costs are given priority before action to facilitating manufacturing. (Sorensson 1978, 5)

In the early 1980s, however, reliability replaced safety as Volvo's top priority. Much of the quality work is conducted by subcontractors to Volvo. The suppliers are completely responsible for the material supplied and for assuring that the parts conform to established specifications. Thus, the supplier is required, by inspection, to verify that product requirements are maintained and to appoint an identifiable person with total responsibility for the quality of the products supplied at a given moment. Furthermore, the supplier must compile a quality handbook, keep documentation, and institute retrieval systems covering the products delivered in those cases where Volvo so specifies.

The quality-control process starts by an initial sample testing to verify that the supplier has correctly understood the specifications. In receiving protocols of inspections, Volvo follows up on the quality of materials and components. Each component used in a Volvo car is classified in one of four categories in order to allocate quality-control resources as effectively as possible. When a component is classified in the category of highest priority, it is marked with a symbol that indicates that the quality specification concerns either a regulation according to law or a safety requirement set by Volvo, or both. This means that the production units must document all the critical steps in manufacturing these items. To exemplify, the producer should document inspection planning, production planning, and material handling. This document is kept on file for 10 years after the quality routines cease to be applicable. The results from the quality inspections are kept five years, along with production and material control records. This is done mainly to demonstrate compliance with specified procedures to national or other authorities. For the other three categories of components, very advanced and powerful statistical decision rules are used for achieving high quality (*cf.* Sorensson n.d.). In fact, the quality-control routines for nonsafety-related Volvo details have more statistical power than the Swedish tests for seat belts.

The Safety-Information Feedback System

The planning, production, marketing, and use of an automobile involves several possibilities for obtaining feedback information for improving the product's safety. In fact, Volvo uses a well-articulated system of extensive feedback for improving the safety of its cars. The discussion to follow analyzes hazard-management and feedback processes according to the format suggested in Figure 4.4.

Hazard-managing agents. The development of a Volvo car from idea to scrapping (i.e., cradle to grave) covers 20 to 30 years (see the central column in Figure 4.4). The median life of a Volvo car in Sweden exceeds 19 years, which is longer than any other car on the Swedish market. Figure 4.4 depicts different stages and opportunities for controlling hazards. Blocking of hazards associated with a car may be initiated or performed by the individual, the company, or society. The individual may block hazards in auto design by selecting a safe model when buying a car, by mounting safety equipment, by making product-liability claims, or by initiating recalls. Society determines the specific and general rules for the individual and for the manufacturer through laws, requirements, and active inspection to assure that products are satisfactory. From the company's point of view, the individual and society define the environment to which the car manufacturer has to adjust, either through adaptation or through efforts to influence its surroundings. Therefore, any description of hazard management in Volvo must be viewed in relation to societal and individual requirements and expectations. All planning of a new car, for example, takes into account available knowledge about the laws that will regulate the safety of the product upon its debut into the market. Likewise, people's future preferences and expectations regarding perceived safety must be estimated and taken into account when designing a new car.

Preproduction process. A changeover to a new car model starts with preliminary ideas as early as 10 years before mass production can begin. The more intensive planning work starts about 5-6 years ahead of mass production of a new car. Plans for changes in existing series of models, however, may occur only 2-3 years ahead of the model change. Because the planning takes a long time, it is important for the manufacturer to anticipate and predict future market conditions and legal requirements.

The planning process runs through different stages and is influenced by many people. To illustrate, a committee is sometimes used to integrate functional characteristics and to bring together experts from different units in the Volvo organization. Thus, crashworthiness is covered by one of the people on the committee, comfort by another one, etc. The committee then follows the construction work through the different prototypes. The functional safety requirements are derived from laws, other regulations accident analyses, and driving and laboratory tests. No explicit formal analyses are made to weigh costs against safety benefits, but both enter into judgments about design.

The initial work in the planning stage is characterized by a comparatively small staff with both formal and informal contacts. Virtually all the individuals working with the development of a new car have their offices and laboratories in the same building. When most intense, the planning work may require a total of more than two thousand people. But in the initiating stage, only about a few hundred or fewer are active. Whereas flexibility is great at the planning stage, the industrial production of automobiles is less flexible because, once in production, the rules for manufacturing the cars become very strict. All changes of a nonsafety-related nature, for example, are performed when the production is pre-planned to undergo changes, usually at the annual model changeover.

Mass production and internal hazard feedback. Independent subcontractors manufacture about 75 percent of the parts of a Volvo car. Volvo's own assembly lines in Canada, Belgium, and Sweden, plus Malaysia, Thailand, and other locations produce all the 240/260 cars. An elaborate quality check system ensures high and even quality. As mentioned earlier, the role of the Safety and Environment Department is that of an adviser, providing legal frames for the final product and proposing routines and ways of checking the products for safety. The safety control work is done locally wherever parts are produced or cars assembled (Figure 4.4).

When the first cars of a model are finished, type approval or self-certification ensures their safety. Every car leaving the factory undergoes final quality

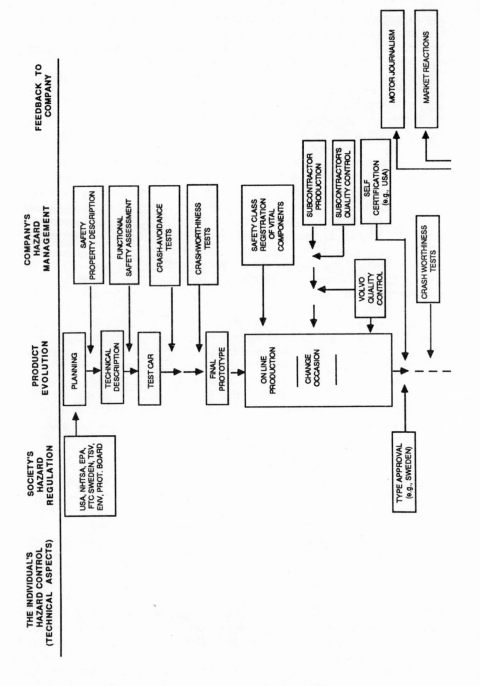

73

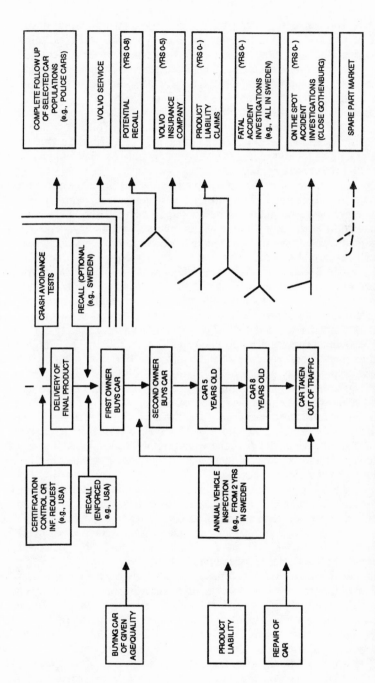

Figure 4.4 Feedbacks enabling improved safety during planning, production, marketing, and use of a car.

control. The company itself performs regular laboratory crashworthiness tests. Of the 70 annual full-scale tests, some use prototypes and test cars, some use regular production cars. In addition, crash-avoidance tests are performed with prototypes, test cars, and regular cars. The characteristics of the regular cars also serve as a baseline in producing prototypes for new models and their crash-avoidance characteristics.

As noted above, certification control is a check of the company's routines made by national authorities to ensure that the company follows the legal requirements in that nation. Officials from abroad may also visit the company and inspect the facilities for controlling the safety and quality routines used by Volvo (e.g., the full-scale crash laboratory).

Car in use. After a car has been delivered to a customer, hazard management changes in character. From mostly an "in-house endeavor," where the company defines and tests almost all safety requirements, safety questions must now be treated in interaction with the public, market, and society. Correspondingly, the feedback to the company becomes slower, more unreliable, and also more directly affected by values other than those of the company.

The extent to which market reactions reflect safety is unclear, but they do in some cases (e.g., the front tank design of the old models of Volkswagen, the Pinto gas tank). The introduction of split brake systems and head restraints may have affected sale volumes for Volvo. Volvo markets its products with safety as an important characteristic and sells its cars to "above average safety-concerned drivers." Motor journalists are important in affecting the market. In general, they tend to concentrate on new cars and neglect the older ones, which is quite unfortunate from a safety point of view. After all, only a small proportion of the cars in traffic are new ones and the change in characteristics of a car over time should be the most interesting topic for the average driver.

Much better feedback comes from the complete follow-ups of some types of cars (e.g., police cars) in Sweden. Because of service contracts for a special fleet of about 1,000 cars per model year, these cars supply feedback over three years. Special repair reports prepared for these vehicles provide information about failures, failure modes, and repair costs.

Volvo service is used when the car is new and the Volvo insurance effective (see below). But when the car gets older, gradually fewer and fewer cars are repaired by Volvo service companies. Most cars five years old or older are not repaired by Volvo service companies any more. Some cars are, however, followed in detail up to 200,000 km (120,000 miles) through special service contracts with the Volvo field service organization.

Product-liability claims. These claims constitute an extremely important type of feedback to Volvo. As mentioned earlier, product-liability claims number only one or two per year and thus initiate few quick changes. But existing laws have been very instrumental in creating routines in the preproduction and production processes to avoid costly product-liability claims. Such feedback is very powerful and stimulates fundamental changes in favor of safety. it is not so quick, however, and it may have less impact in industrial safety upon a leader (such as Volvo) than upon other corporations.

Volvo Insurance Company. Volvo is unique in the automobile industry in having a long series of detailed information about its cars during the first five years because the sales contracts have included free warranty in a Volvo-owned insurance company (Volvia). A company may wish to analyze the costs of increased safety (e.g., the safety improvements obtained by constructing longer deformation zones), but this type of information cannot be obtained from outside the company. This feedback source, by contrast, is highly reliable, quick, and confidential within the company. The feedback works reliably, however, only during the car's first five years. After this, many owners do not renew their Volvo Insurance Company contracts covering repair costs.

Recalls. Information about potential recalls may not lead to a recall but still provides data of value for safety engineering. Recalls have, up to recently, been optional in Sweden and mandatory in the United States during the first eight years of the car.

Accident analysis. The accident investigation group registers all Volvo accidents in Sweden. If any of these accidents involve fatalities, the accident is analyzed and the result put on file. Fewer than 100 such accidents occur in Sweden each year. This feedback is not totally reliable since second-hand information (e.g., police and insurance reports) has to be used in most of the cases. The feedback is also rather slow but it has the advantage of covering the whole life of a car.

As mentioned earlier, the Accident Analysis Group performs an on-the-spot investigation when an accident occurs within an hour's drive from the Volvo Safety Center. The feedback obtained here is direct and provides evidence about hazardous details not possible to identify in the laboratory. Such investigations affected, for example, the design of the Volvo steering wheel. Crash tests with dummies had not revealed the disadvantages of a steering wheel with only two spokes attaching it to the center axis. Now, four spokes allow a better distribution of the force on the body in case of a collision.

The on-site accident analyses are not statistically representative, but the information is of high quality and validity since it is obtained in real traffic and is analyzed directly. Therefore, this feedback may have speedy and significant effects on the preproduction and production processes. Volvo's (as well as most other) accident investigations, however, seem to concentrate on the crashworthiness aspects. Therefore, precrash safety must be thoroughly treated in other accident "on-the-spot" investigations or in other stages (e.g., in handling tests) of the auto life-cycle.

The spare-parts market. The safety spare parts market provides rather slow and unreliable information because so many parts are not manufactured by Volvo. During the first years, however, the information is better. Unfortunately, as the car becomes older and presumably less safe, this feedback gets weaker and weaker.

Annual vehicle inspection. The annual vehicle inspection (depicted on the left of Figure 4.4) provides excellent feedback concerning a number of safety-related details. It seems to be the only reliable feedback source that works throughout the lifetime of a car, and the inspection is thus important for ensuring a high level of technical safety of the vehicles on the roads and for providing valuable information for the manufacturer.

Summary. In summary, the information feedback system for safety information to the company is extensive and valuable for safety assurance. All together, company and society feedback loops provide an information network of a high quality which the company utilizes efficiently. Some of the feedback sources are more general, including safety aspects (e.g., complete follow-ups and the Volvo insurance company) whereas others are primarily safety-oriented (e.g., fatal-accident investigations and annual vehicle inspections). Some of the feedback loops are managed internally by Volvo and are open for external researchers to a greater (e.g., on-the-spot accident investigations) or less (e.g., insurance data) extent. The information on the external feedback from the annual inspections is public and available to any interested party for further analyses.

Conclusions

Corporate management of product hazards, it is clear, is a process in which a producer, societal representatives, and the public interact in seeking a balance among the benefits, costs, and risks of a product. From this analysis of Volvo, high-quality hazard management appears to be characterized by:

- easy access for all parties to reliable information about a product's benefits, costs, and negative consequences (i.e., deaths, injuries, or illness),
- a good understanding of the relation between changes in the product or its use and changes in the safety level (as well as good knowledge of the engineering skills needed to achieve changes that may be requested),
- the existence of an information network that informs the interested parties about changes in the safety level of new products and products in use,
- an appropriate organizational structure in society and the producer for negotiating and deciding on a justifiable balance among benefits, costs, and risks as experienced by different parties,
- an appropriate societal control over incentives and penalties for the producer and users to achieve the determined level of safety.

Viewed by these criteria, Volvo's accident hazard management provides an excellent example of successful corporate hazard management. The information about accidents (e.g., the number of deaths in accidents) is readily available and often of a high quality. The effects of changing characteristics of the automobile or its use have enjoyed extensive study and Volvo possesses the necessary engineering skill. Further, Volvo has access to an impressive information and feedback network for learning about the safety of its cars (see Figure 4.4). In Sweden, unlike the United States, negotiations between society and industry characteristically involve cooperative interplay rather than antagonistic confrontation. In the case of automobile accident hazards, the result in Sweden is an agreement on a very high level of traffic safety. The societal technical control carried out by the Swedish Testing Agency, by Volvo, and in foreign inspections is quite strict and contributes to a high level of safety in reality as well as in theory.

Historically, at least five sets of important factors kindled and kept alive Volvo's strong tradition of accident hazard management. The first set of factors includes the general quality and safety-minded attitude of the technicians working with Volvo cars (*cf.* Ingelstam 1980 who argues that it may be just as interesting to investigate the social psychology of the staff of a corporation as its inferred decisions). In the automobile manufacturing industry in general, the existence of prototype safety cars seems to have been very important because they created targets for improving the safety design of mass-produced cars. The second set of factors includes the view of safety also as economically positive; in short, safety can sell Volvo cars. A third set of factors reinforcing the safety work already initiated by Volvo consists of United States regulations (particularly concerning product liability and recall) introduced in the 1970s. A fourth set of factors is the general societal, public, and media interest in safety issues, without which a successful hazard-management process would have been impossible. A fifth factor was the establishment in the 1960s of the Swedish Motor Vehicle Inspection Company, which prompted safety engineering and repairs.

Viewed in this perspective, the Swedish societal authorities have not played the central role in pushing for the technical safety standards of Swedish cars. Whatever the true story about the process leading to societal regulation, the results in terms of accident safety were quite satisfactory and the standards set in Sweden among the strictest in the world. Today, however, the authorities often express their opinion on safety regulations in paraphrases of the "importance of international harmonization in standard settings in the field of motor vehicles." This simply means that it seems feasible to change the Swedish safety standards so that they become more similar to other European standards. Such changes are good for the manufacturers who can sell the same car all over the world, but they carry negative aspects as well. Climate or culture may be important reasons for having one set of regulations in one part of the world and another set of regulations in another part (e.g., corrosion, daytime lights). The internationalization of safety-related requirements does not make such adaptation possible. Eventually, one solution will dominate all the others if the trend continues. Another and more important issue is that international regulations may be based on one-shot erroneous research or worldwide known but biased information.

There is still room, of course, for safety improvement at Volvo and in the automobile industry. More industrial and societal attention should focus specifically on the safety of the aging car. Currently, the annual inspection carries most of the responsibility for the safety of the old car with its different types of defects due to wear. From a general engineering point of view, society may benefit from more attention to crash-avoidance properties and accident statistics. Performance demands (as contrasted with the manufacturer's product design) should be required for new and old cars and society should implement regulations concerning the accessibility of different parts for standardized and easy checking, as with the case of airplanes.

ACKNOWLEDGMENT

This study was supported by a grant from the Swedish Council for Research in the Humanities and Social Sciences and grants to Decision Research, A Branch of Perceptronics. The author is greatly indebted to may individuals in and outside of Volvo for providing important information and wants to thank them all for their kind help. In particular, he wishes to thank Lennart Strandberg for his support in the initial stages of this project and Jeanne and Roger Kasperson for their careful editing of the manuscript.

5

Managing Occupational and Catastrophic Hazards at the Rocky Flats Nuclear Weapons Plant

Operating since 1951, the Rocky Flats nuclear weapons plant near Denver currently employs 3,700 people, has an estimated value of $2 billion, and supports a total floor space of 2.1 million square feet. Though purely a military facility, the plant is financed by the Department of Energy and is operated for the government by Rockwell International, Inc., a major defense contractor.

The plant itself occupies 384 acres at the center of an 11-square mile exclusion zone and is guarded by armed sentries and a 12-foot-high security fence topped by barbed wire. The exclusion zone and the area beyond are a rural oasis in one of the most rapidly growing U.S. metropolitan areas. Abutting the zone are cattle ranches and a few scattered houses. No public facilities or institutions such as schools, prisons, or hospitals lie within five miles of the plant, and only 5000 people reside within that radius. Yet within 10 miles one finds over 250,000 residents, and within 50 miles 2.5 million. By the year 2000, the population within 10 miles is projected to reach 400,000 and that within 50 miles 3.5 million.

The original mission of Rocky Flats was the manufacture of nuclear weapons components made of plutonium, enriched uranium, depleted uranium, and conventional metals. Enriched uranium processing was later discontinued, whereas beryllium processing was added. With the growth and aging of the U.S. weapons stockpile, estimated at 26,000 warheads in 1982 (Arkin, Cochran, and Hoenig 1982), Rocky Flats acquired the mission of disassembling and reprocessing weapons components that it had originally produced.

Today, the Rocky Flats plant receives new plutonium from production facilities at Hanford, Washington, and Savannah River, Georgia. It also receives old plutonium from obsolete warheads that have been dismantled elsewhere. The old plutonium is chemically purified by removal of americium-241 contamination, and both old and new are chemically transformed into metal, machined into

79

new components, and shipped for assembly to the Pantex Plant near Amarillo, Texas. Radioactive wastes produced through chemical treatment are sent to federal waste repositories. Though Rocky Flats never sees an assembled nuclear warhead, it is the sole clearing house for all U.S. weapons-grade plutonium. As such it will contribute to the production of perhaps 37,000 new warheads between 1982 and the mid-1990s (Table 5.1).

This chapter, which draws heavily on an earlier analysis by the principal author (Hohenemser 1987), reviews risk assessment and management at Rocky Flats from several perspectives. First, it recounts the controversies that have surrounded the plant since its founding in 1951. Second, it offers a descriptive tour of the plant, with emphasis on hazard management strategy. Finally, an analysis of occupational health issues and a discussion of the management of low-probability/high-consequence accidents conclude the chapter.

A History of Controversy

Unlike the other cases treated in this volume, the Rocky Flats plant has been the subject of controversy almost from its inception in 1951. On several occasions, fabrication procedures resulted in plutonium fires that burned through the high-efficiency particulate air (HEPA) filters (Table 5.1). Because plutonium, a potent carcinogen, is hazardous to human health in microgram quantities when inhaled, the plutonium fires at the Rocky Flats have aroused widespread public fears, not only through the events themselves but also through the lack of candor displayed by plant officials in explaining them.

In recent years, Rocky Flats became the target of repeated large and small demonstrations organized by local and national groups. The largest protests have come from peace groups who object to nuclear weapons in principle and want their production halted. Less vocal protests have come from supporters of nuclear weapons, local environmentalists, and landowners who consider the present location of the plant too hazardous.

In July 1985, landowners whose land abuts Rocky Flats settled three long-standing lawsuits that alleged approximately $20 million in damages due to plutonium contamination. Under the out-of-court agreement, the plaintiffs released 475 acres to Jefferson County and 368 acres to the city of Broomfield. In return, Rockwell International paid $8.7 million from Department of Energy funds, Jefferson County contributed $1.15 million, and Broomfield $894,000. The plaintiffs retained 1200 acres, but Rockwell paid $150,000, for remedial action to assure that the property would be suitable for commercial development. Interestingly, the settlement of lawsuits occurred despite Judge Richard Matsch's finding of "no scientific basis" for concluding that soil and air concentrations of plutonium and americium on the properties produce health effects different from background radiation.

Founded in nearly total secrecy in 1951, the Rocky Flats plant has more recently generated public concerns that have directly or indirectly produced a series of studies and reports that evaluate the plant's impact on the local area. One major study, the three-volume *Final environmental impact statement* (DOE 1980), provides data on the plant, its fire experiences, its occupational health

Table 5.1

MAJOR ACCIDENTAL RELEASES OF PLUTONIUM FROM THE ROCKY FLATS PLANT

YEAR	ACCIDENT DESCRIPTION	TOTAL PLUTONIUM INVOLVED IN PLANT	TOTAL OFF-SITE RELEASE[c]	
		kilograms	microcuries	grams
1957	Plutonium fire in production building	250[a]	25618	0.05
1958-68	Contaminated oil leakage from storage drums	--	3.4×10^6	7
1964	Chemical explosion in glove box	----	10	2×10^{-5}
1965	Glove box drain plug fire	?	1170	2×10^{-3}
1969	Plutonium glove box fire and building fire in production building	1,000[b]	856	2×10^{-3}
1969	Plutonium fire in tunnel between buildings	?	20	4×10^{-4}
1974	Release from control valve failure	----	934	2×10^{-3}

[a] Based on filter loading of 620 filter main filter plenum, as discussed by Johnson (1981). The entire main building plenum burned through during the 1957 fire. A total of 14-20 kilograms of plutonium was unaccounted for.

[b] Based on estimates by E. W. Bean, Attachment B to July 2, 1976 letter to J. F. Burke, "Cost of May 11, 1969 Fire at Rocky Flats," as quoted by Chinn (1981).

[c] Release estimates given in *Final environmental impact statement* (DOE 1980, vol. 1, p. 2-169).

record, and its capacity to resist natural events such as high wind and earthquakes. A second major study, the *Long-range Rocky Flats utilization study* (DOE 1983), originated in a request by a state-created 1974 citizens' task force, which asked that the government consider converting Rocky Flats to a less hazardous energy-related research facility and moving plutonium work elsewhere.

The experience of the principal author (Hohenemser) of this chapter came through service as one of 13 technical advisors to a citizens' task force appointed by Governor Lamm and Congressman Wirth in 1981. Known as the Blue Ribbon Citizen's Committee, the group was charged with oversight of the Department of Energy's *Long-range Rocky Flats utilization study*. During 1981-83, the committee members toured the plutonium processing areas, received drafts of study elements, and were invited to comment on the accuracy and completeness of the DOE study. The technical advisors also reviewed the Blue Ribbon Citizen's Committee's own report (BRCC 1983) on the DOE study.

The *Long-range Rocky Flats utilization study* and the accompanying Citizen's Committee report provided a new "citizen validated" assessment of the Rocky Flats operations in the context of relocation costs. This assessment sidestepped the dangers of nuclear weapons *per se* but did address issues of occupational and public health and safety, catastrophic accidents, and eventual reconversion of the plant. According to the *Long-range Rocky Flats utilization study*, the greatest hazard at Rocky Flats is plutonium dispersal initiated by natural events, with 87 percent of public health risk attributed to potential impact of high wind on plutonium buildings and an additional 6 percent to earthquakes. Recently revised safety analyses reverse the relative contributions of these two natural phenomena, with earthquakes now dominating at 66 percent and severe winds dropping to 33 percent of the composite risk (GAO 1987, 40). In numerical terms, plutonium accidents are expected to produce a cancer risk equal to 1 in 900 million per year for individuals living within 50 miles of Rocky Flats. This number appears entirely negligible when compared to other risks that individuals face. It also seems minor when compared to estimated relocation costs of $4 billion (in fiscal 1986 dollars; *see* GAO 1987, 26).

A recent controversy at Rocky Flats has been public debate over a trial burn of radioactive mixed waste in an incinerator. Although the Centers for Disease Control estimated that the risk that a Colorado resident would contract a disease from an accidental release was one chance in 68 quadrillion, the test series of six burns has been delayed several times due to the objections of environmentalists (*Nuclear Waste News* 1987, 280).

For a plant that produces the most lethal products in the world using one of the most toxic substances known, Rocky Flats appears to have maintained a remarkably low level of risk. How is it that risk can be kept so low? The answer to this may carry lessons for the corporate management of health and safety risks more generally.

A Tour of the Plant

Rocky Flats is no ordinary industrial plant. It lives and breathes hazards. A rough estimate of the plutonium handled per year is 5-10 tons. If plutonium were steel and Rocky Flats an ordinary foundry and machine shop, the facility might need three or four workers, a boss, a secretary, a furnace or two, eight to ten machines, and perhaps 2,000-4,000 square feet of floor space. Because plutonium is not steel, but a potent, radioactive carcinogen that is flammable and capable of chain-reaction accidents and diversion for weapons purposes, Rocky Flats requires a workforce and space that are about a thousand times greater.

As one approaches the plant, the first sign of hazards-to-come is a notice on the access road reminding one that the current record of accident-free days is some large number. Then come the security checks by the armed guards at the main gate, where a number of military vehicles, including two army tanks, are visible. Driving around the drab "campus" of Rocky Flats, one sees cement-lined solar evaporation ponds used to recycle process water without contaminating the local surface and ground water. Entry to the seminar room for initial briefing brings further security checks by armed guards, and even trips to the toilet are supervised by security personnel.

The centerpiece of the plant is the new plutonium recovery and waste-treatment complex. Designed in 1974 to replace an older building with one providing a higher standard of radiation protection, this 331,000-square-foot facility, with two floors above and eight floors below ground level, cost $300 million and is by any measure a monument to engineered safety. To appreciate the hazard control strategy used, it is necessary to understand why plutonium is a threat.

A dense grey metal that is warm to touch, plutonium in air quickly becomes covered with a dull-green surface oxide having the consistency of a fine powder. This oxide is loosely attached to the metal and easily adheres to surfaces with which it comes into contact. In finely divided forms, plutonium can ignite spontaneously. The plutonium used at Rocky Flats emits ionizing radiation from five isotopes, the most abundant of which is plutonium-239. The primary threat of handling plutonium is the possibility that finely divided oxide particles are inhaled, or that workers receive cuts through which finely divided plutonium debris enters the body.

A few millionths of a gram of plutonium are sufficient to pose a cancer threat to humans. Accordingly, the Department of Energy sets the maximum permissible body burden at 40 nanocuries, or 0.55 millionth of a gram, with bone and lung considered the most susceptible organs (DOE 1980, p. G-3-1 ff). Body burdens are regularly monitored for Rocky Flats workers, using whole body counters that are sensitive to the low energy gamma rays emitted by the plutonium and its radioactive decay product, americium. As of 1980, Rocky Flats employed 14 individuals who had received more than 50 percent of the maximum permissible body burden (Putzier 1981). Workers with open wounds do not work in plutonium areas, where no smoking, eating, or drinking is allowed; plutonium workers must wear protective clothing and carry respirators for emergency use.

The threat of airborne plutonium extends beyond the workers to the general public, since respirable oxide particles are easily transported long distances in air. The principal engineering controls used to protect workers and the public are inert atmospheres to block oxidation and containment and filtering to prevent dispersion of particulates.

Escorted into the deep interior of the new plutonium building, one eventually enters the rooms in which plutonium is processed. The guide explains that though not in use yet, these facilities are similar to others at Rocky Flats. What one sees are stainless steel plutonium enclosures, known as "glove boxes," that contain an inert atmosphere to prevent oxide formation. Rubber, lead-impregnated gloves extend into the interior and permit workers to manipulate tools and materials. The lead in the gloves and lead-glass windows protect against the gamma and beta rays emitted by the plutonium. Automatically triggered radiation and fire warning systems signal accidents should they occur.

Administrative controls and limitations on size of the containers assure that plutonium quantities do not reach levels at which chain reaction accidents can occur. Known as criticality accidents, these are precursors of nuclear explosions in which plutonium undergoes a rapidly increasing reaction leading to fission of a small fraction of the plutonium present. Criticality accidents generate some heat and a very high flux of deadly, high-energy neutrons. So far in its 35-year history, Rocky Flats has reported no criticality accidents, though such accidents occurred during the Manhattan Project in World War II and at other nuclear weapons laboratories in the United States since then.

The interiors of glove boxes are interconnected by conveyer lines, permitting movement of plutonium from one operation to another without ever bringing the material into contact with the room atmosphere. Glove boxes are grouped in clusters that are divided from other clusters by wide corridors and fire doors. Air locks separate the plutonium-processing rooms as a whole from the rest of the building. There are three atmospheric zones: Zone 1, inside the glove boxes themselves; zone 2, comprising the rooms in which the glove box clusters are located; and zone 3, comprising the rest of the building. The ventilation air pressure of zone 1 is negative with respect to that of zone 2, and zone 2 pressure is negative with respect to zone 3. Air passes, if at all, from nonradioactive areas in zone 3 to areas with increasingly more radioactive environments, with final exhaust from zone 2 or 1 through four-stage HEPA filters. As an added precaution, the stack effluent is continuously monitored for radiation.

At one point in the tour of the new building, one passes a storage vault for plutonium and learns that it contains fabricated metal pieces and canisters of plutonium nitrate solution received from Hanford and Savannah River. The vault has an inert atmosphere enclosed by thick concrete walls. Operators handle samples through a three-axis retriever-robot, which runs along a central track down a long corridor and can lift samples from one-meter concrete cubicles on each wall of the corridor. The walls of the corridor are approximately four stories high and a football field in length. Required separation of the plutonium samples by one meter is designed to prevent criticality accidents.

The key to the Rocky Flats ventilation system is the use of four-stage HEPA filters. So important are these filters that the plant maintains a separate unit, including a 15-person work crew that is permanently assigned to filter testing and maintenance. HEPA filters consist of densely packed fiberglass mounted into tightly sealed, corrosion-resistant frames. Instrumentation inside the filter plenum monitors throughput and radiation buildup. The twofold purpose of the filters is (1) to block routine emissions of finely divided plutonium particulates, and (2) to ensure that even during a serious accident significant quantities of plutonium will not be released to the environment.

The"design-basis accident" for Rocky Flats HEPA filters is the inadvertent burning of 500 grams of plutonium in a glove-box fire. During subsequent cleanup, this is estimated to produce a 12-gram-per-day particulate flow at the input of the filter system. Under these conditions the decontamination factor provided by a four-stage HEPA filter plenum is expected to be about one trillion. Assuming that all particles are in the filterable size range, this suggests that of a trillion incident particles, only one on the average would emerge on the ambient air side of the filter. Filters may transmit significant fractions of particulates below 0.3 micrometer in diameter, and therein lies their weakness.

On at least two occasions, the filters have failed because plutonium fires burned through the filter plenum and released plutonium particulates into the environment. According to information provided by the Rocky Flats Plant (DOE 1980), fires in 1957 and 1969 released about 50 milligrams and 400 micrograms of plutonium, respectively. As small as these quantities appear in absolute terms, they must be understood with reference to the minute quantities of plutonium needed to present a threat to humans. In addition the amount of plutonium burned in the two fires far exceeded the quantity (i.e., 500 grams), assumed to be involved in a glove-box design basis accident. It is estimated that as much as 250 kilograms of plutonium may have been lodged in the 620 filters destroyed by the 1957 fire and that as much as 1000 kilograms of plutonium were oxidized in the 1969 fire (Table 5.1).

Rocky Flats experienced a serious accident during the period 1959-1969 when waste cutting oil containing an estimated 1.5 grams of plutonium leaked into the ground from 1000 steel storage drums. Of this amount about 80 percent stayed on site, and 20 percent found its way off site via wind dispersion (DOE 1980). According to the Department of Energy, the leaking oil drums produced 99 percent of all plutonium that ever got off site and are responsible for the bulk of the airborne plutonium in the environs of Rocky Flats. Critics suggest that additional environmental deposits may have originated in the 1957 and 1969 plutonium fires (Johnson 1981, Chinn 1981).

The Colorado Department of Public Health, the Department of Energy's Environmental Measurement Laboratory (Krey 1974, 1976; Krey and Hardy 1971), and the Colorado Committee for Environmental Information (Poet and Martell 1972) have all monitored the soil content of plutonium on lands surrounding the plant. In general, the various measurements have yielded similar results for surface plutonium levels but have not answered the question of how much plutonium has been released. The contour map in Figure 5.1 depicts soil plutonium levels around Rocky Flats, as prepared by Krey and Hardy (1971).

Finally, theft of plutonium and diversion for weapons purposes by unauthorized individuals is an ever-present threat. In handling quantities of plutonium of several million grams per year, Rocky Flats seeks to account for the material at the one-gram level, employing extensive, computerized materials accounting methods. It is very difficult, however, to succeed at that level of

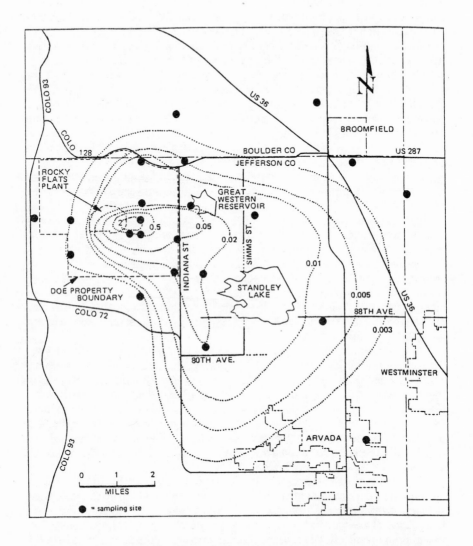

Figure 5.1   Plutonium-239 contours around Rocky Flats in microcuries per square meter.

SOURCE: DOE (1980, p. 2-74).

accuracy. Complete accounting, for example, occurs only a considerable time after a successful thief would have left the plant. For these reasons, materials accounting must be supplemented by other methods, such as security checks. Even these are limited, however, since exit radiation counters that monitor departing employees cannot detect samples of plutonium for which tell-tale gamma radiation has been shielded by relatively lightweight material.

The plutonium hazards at Rocky Flats are therefore challenging and complex, more so than other industrial hazards treated in this volume—even PRODUCT-A at PHARMACHEM (chapter 3) and methyl isocyanate at Bhopal (chapter 6). The management of plutonium hazards enlists a classical strategy of "defense in depth," in which administrative controls, engineered safety, personal protection, and warning and monitoring are in some cases stacked eleven and twelve deep in order to avert hazard consequences (Table 5.2). This strategy is particularly apparent in the management of worker and public protection against internal exposure from airborne particulates. A significant aspect of defense in depth is that many controls that are essential for worker protection also protect the general public (Figure 5.2).

Further evaluation of hazard management at Rocky Flats entails an examination of two major issues: (1) the occupational health record, as embodied in the health statistics of more than 10,000 employees who have worked at Rocky Flats since 1951; and (2) the assessment of potential catastrophic risks arising from events which to date have *not* occurred at Rocky Flats.

Occupational Health at Rocky Flats

One aspect of occupational health and safety is worker accident experience. The most broadly applicable measures of this are two indices—frequency and severity—developed by the National Safety Council. Measured on these scales, the Rocky Flats workforce has an enviable record. Its accident frequency index has run at approximately 3 percent and its severity index at 10 percent of the all-industry average (DOE 1980), much lower than even the good experience at PETROCHEM and PHARMACHEM (chapters 2 and 3). The only significant exception is the year 1967, when the severity index exceeded the all-industry average.

A second important measure of occupational health at Rocky Flats is external radiation exposure, as determined by film badges worn by nearly all employees. In 1980, 1.6 percent of 3,870 Rocky Flats employees received whole-body external exposure of greater than 1 rem. Looking back in time, one finds that this low rate has not always prevailed (Figure 5.3). In 1966, of 3,175 badged personnel 28 percent had exposures in the excess of 1 rem, and 2.7 percent exceeded today's occupational exposure standard of 5 rem per year. In 1967 Rocky Flats voluntarily adopted the 5-rem annual standard before it became accepted nationally, and since then external radiation exposure has dropped sharply. A 1981 study for Rockwell International reported that Rocky Flats could meet a 2.5-rem standard with no additional employees and a 1-rem annual standard with only 97 additional employees (Putzier 1981). Exposure reduction

Table 5.2

**SOME PLUTONIUM HAZARDS AND HOW THEY ARE CONTROLLED
AT ROCKY FLATS**

HAZARD	HAZARD CONTROLS
Plutonium air particulates: dispersal and worker exposure	1. Administrative control 2. Glove box containment 3. Inert atmosphere 4. Ventilation air pressure zones 5. HEPA filters 6. Radiation warning systems 7. Fire warning systems 8. Fire protection systems 9. Glove box cluster isolation 10. Protective clothing 11. Emergency personal respirators
Plutonium air particulates: dispersal and public exposure	1. Administrative control 2. Glove box containment 3. Inert atmosphere 4. Ventilation air pressure zones 5. HEPA filters 6. Radiation warning systems 7. Fire Warning systems 8. Fire protection system 9. Glove box cluster isolation 10. Stack gas monitors 11. Public exclusion zone 12. Soil monitoring
Plutonium criticality: worker neutron exposure	1. Administrative control 2. Limited size of containers 3. Restricted geometry storage 4. Radiation warning systems 5. Neutron shields
Plutonium: worker gamma ray exposure	1. Administrative control 2. Physical shielding 3. Personal radiation monitors
Plutonium theft: clandestine weapons	1. Administrative control 2. Plutonium accounting procedures 3. Security control 4. Exit radiation monitors
Plutonium bearing oil: dispersal and public exposure	1. Administrative control 2. Barrel storage 3. (Permanent disposal)
Plutonium bearing water: dispersal and public exposure	1. Administrative control 2. Evaporation ponds 3. (Permanent disposal of sludge)

89

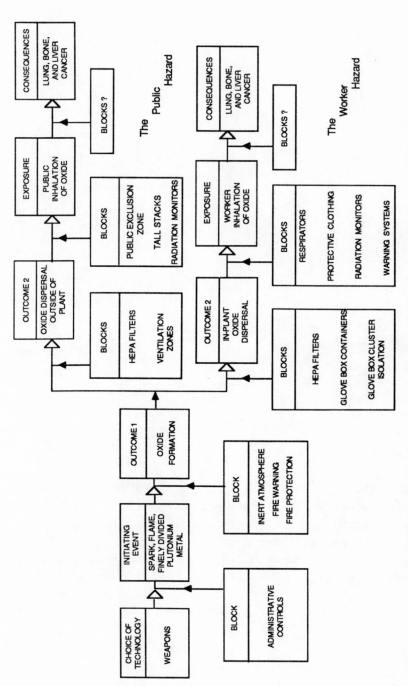

Figure 5.2 A diagram of the causal structure of hazard, illustrating public and worker hazard and controls employed.

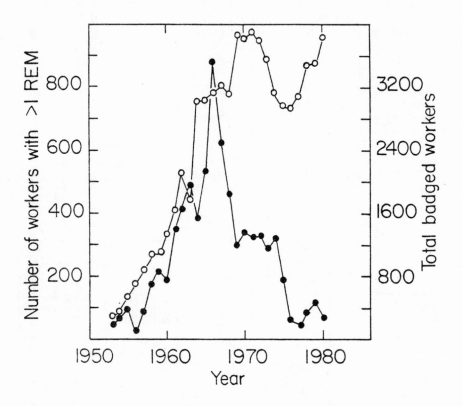

Figure 5.3 **Whole-body external radiation exposure among Rocky Flats workers since 1953. The solid points show the number who received external exposures greater than one rem in each year. The open points show the total badged workforce.** (Data from Putzier 1981)

since 1966 is not simply a case of "spreading the dose" to more individuals but is the result of a significant drop in total exposure.

For the average worker, the annual level of radiation exposure at Rocky Flats exceeds background levels due to all other causes, such as natural background and medical x-rays, by a factor of two to five. In extreme cases, when exposure approaches the 5-rem annual standard, occupational exposure exceeds background by a factor of 30. The health effects of external radiation exposure may be estimated by using dose-effect relations recommended in the so-called "BEIR report" by the National Research Council's Committee on the Biological Effects of Ionizing Radiation (National Research Council 1980). This leads to the estimate that as many as 6 cancer mortalities and 9 genetic effects may occur among all workers employed at the plant since 1951. These effects must be compared to more than 2000 cancer mortalities expected for the same workers from causes other than occupational exposure to external radiation.

A third measure of worker health at Rocky Flats is the record of internally accumulated plutonium, which is monitored through urine analysis and whole-body radiation scans utilizing low-energy gamma emissions of plutonium and americium. In 1980, 22.6 percent of Rocky Flats badged personnel had measurable systemic accumulations of plutonium. Though the bulk of these burdens fell in the range of 3-10 percent of the maximum permissible amount of 40 nanocuries (nCi), 14 workers exceeded the 50 percent level at which Rocky Flats management cuts off all further plutonium work (see Table 5.3). The 1,659 present and past Rocky Flats workers with plutonium burdens carried a total of about 6000 nCi. Using standard organ-dose models, calculations lead to an estimated 9-15 additional cancer mortalities over the lifetime of the Rocky Flats workforce.

The dose-response models on which such estimates rely are considerably more problematical than the corresponding data for external exposure. Hence an alternative to their use is to conduct direct epidemiological studies on Rocky Flats workers. Since these workers constitute about one-third of all U.S. workers with measurable plutonium burdens, and many are now entering an age when elevated cancer mortality should become detectable, the Rocky Flats workforce offers the best available chance for detecting such effects if they exist.

The first results of such epidemiological studies appeared in reports by George Voelz and collaborators at the Los Alamos Scientific Laboratory in 1981 (Voelz *et al*. 1981). Voelz's group worked with an advisory board of seven distinguished epidemiologists and concentrated on mortality data for 7,112 white males who worked at Rocky Flats during 1952-79. The 452 observed deaths from all causes were significantly fewer than the 831 expected in an equivalent group drawn fro the U.S. population as a whole. The 107 deaths due to cancer of all kinds were also significantly fewer than the 167 expected from these diseases. The comparison to the U.S. population as a whole is probably inappropriate for a group of highly paid, highly educated technical and scientific personnel. Voelz recognized this fact and identified it as a possible example of the "healthy worker syndrome" (i.e., all regularly employed worker groups are healthier than their counterparts in the general population) that regularly confounds poorly controlled occupational epidemiology.

Table 5.3

INTERNAL BODY BURDEN OF PLUTONIUM
AT ROCKY FLATS

FRACTION OF MAXIMUM PERMISSIBLE BURDEN (percent)	ACTIVE WORK FORCE IN 1980		PAST AND PRESENT WORKFORCE IN 1980 (number)
	(number)	(percent)	
3-10	640	16.4	1220
10-20	190	4.8	350
25-50	35	0.9	62
>50	14	0.3	27
TOTALS	879	22.4%	1659

SOURCE: Data from Putzier (1981).

Rockwell International, the managers of the Rocky Flats Plant, seized upon Voelz's results as "good news." In a press release in October 1981, Rockwell described all aspects of the Voelz study except its confounding healthy-worker syndrome.

Critics of the Rocky Flats Plant were outraged by the circumstances surrounding the release of the Voelz study. George Voelz was hurt (*Denver Post* 1983) and noted that he clearly described the healthy-worker syndrome in his original paper. And the general public was, as frequently occurs, confused by the highly polarized, rancorous public discussion that followed. After the dust cleared, about the only thing that one could conclude from Voelz's 1981 study is that plutonium-induced cancers are so few that the general good health of the Rocky Flats workforce easily overwhelms them.

Since the publication of the Voelz study, two other increasingly sophisticated epidemiological studies have appeared. One considers the fact that some of

Voelz's cohort was "lost to follow-up" (Acquavella *et al.* 1982). Another study (Wilkenson *et al.* 1983) takes the healthy-worker syndrome as starting point and compares Rocky Flats workers with and without internal plutonium exposure. A third study (Reyes *et al.* 1984) focuses on relatively high incidence of brain cancer, which Voelz had already noted. Unfortunately, the need to make intra-cohort comparisons strongly limits the statistical power of the more recent work, and one is left with the uncomfortable result that the search for pluto-nium-induced health effects in Rocky Flats workers is still inconclusive. A detailed analysis leading to this essential point appears in the *Report of the Rocky Flats employees health assessment group* of the Colorado Governor's Science and Technology Advisory Council, headed by Robert Lawrence (1984).

The occupational health situation at Rocky Flats thus presents some inter-esting contrasts. On the one hand, conventional accidents are at a low level, average and total external radiation exposure is declining, and epidemiological studies of workers have provided no direct evidence for radiation induced health effects. On the other hand, efforts to quiet doubts about potential effects of internal exposure from worker plutonium burdens appear doomed, both because of the arguable nature of internal dose models as well as the statistical limits of direct epidemiological analysis. Moreover, a recent safety appraisal (DOE 1988) highlights inadequate attention on the part of management, to radiological pro-tection.

Assessing and Managing Catastrophic Hazards

Because of the extreme toxicity of plutonium, it is possible to envision events that trigger breaches of containment that lead to disastrous consequences. With the possible exception of the two major plant fires already mentioned (Table 5.1), no such events have occurred in the 35-year history of Rocky Flats. The potential triggering events may be both natural and technological in origin. According to the 1980 *Final environmental impact statement* (DOE 1980), the most damaging of these rare events would be the crash of a fully fuelled, wide-body jet into the plutonium building. According to Department of Energy estimates (DOE 1980), the event has a probability of one in 10 million per year and is expected to release about 100 grams of airborne plutonium and cause 100-300 cancer mortalities among members of the general public. Other rare events with serious consequences include earthquakes, high wind, and a series of pos-sible operational accidents involving explosions and fires.

It is, therefore, not surprising that the study of rare events, their conse-quences, and associated cancer mortality risks was a major component of the *Long-range Rocky Flats utilization study* and the work of the Blue Ribbon Citizen's Committee. Structurally, the problem is similar to the evaluation of nuclear reactor accidents, or the chemical release that occurred at Bhopal (chapter 6), which involve massive, normally contained inventories of toxic materials that can potentially produce large numbers of fatalities and genetic effects. Rocky Flats differs from nuclear reactors in two important respects. First, with the possible exception of jumbo jet crashes, Rocky Flats lacks a large energy source, comparable to a reactor core melt, that can trigger containment breaches.

Secondly, Rocky Flats is considerably more diverse and complex than a nuclear power plant and has far more employees. Taken together, these characteristics should lead at Rocky Flats to less severe "maximum" disasters but to more frequent smaller events.

Prior to the *Long-range Rocky Flats utilization study*, the most thorough treatment of rare events had appeared in the *Final environmental impact statement* (DOE 1980). The method involved probabilistic risk analysis (PRA) consisting of four steps:

1.   Selection of several scenarios, each triggered by an extreme initiating event (e.g., high wind, mechanical failure, earthquake, etc.), involving the release of plutonium.
2.   Estimation of the probability that a given scenario should occur, up to and including the release of harmful material.
3.   Determination of the radiation dose delivered to members of the public.
4.   Calculation, employing standard dose-response models, of the expected cancer mortality.

This methodology closely parallels research conducted on nuclear power plants (NRC 1975). As applied here, however, it differs in one important respect. Whereas the 1975 *Reactor safety study* made the claim of having considered *all possible significant* scenarios, the Rocky Flats analysis deliberately considered only two scenarios for each type of initiating event: (a) a "maximum credible accident"; and (b) a "maximum probable event." Translated into English, this means that analysts made a judgment about what constitutes the "worst" scenario having a probability greater than one in a million and called this the "maximum credible"; and, similarly, analysts considered the worst scenario having a probability greater than one in a thousand and called this "the maximum probable". By analyzing just two events one may gain an idea of the range of consequences that could occur without incurring the high cost of a fuller analysis. At the same time, this approach precludes a quantitative statement of total risk.

What finally emerged from the $1.15 million safety and risk analysis component of the *Long-range Rocky Flats utilization study* was an assessment that paralleled earlier work under the *Final environmental impact statement* but added the full risk analysis methodology pioneered by the *Reactor safety study*, or WASH-1400 (NRC 1975). Maximum credible and maximum probable events were analyzed for six buildings and six classes of initiating events, including operational accidents, external missiles, aircraft crashes, earthquakes, high winds, and tornadoes. The risks for these events, expressed in total cancer mortality among the public, were added (Table 5.4) to obtain an annual total risk of 0.002 cancers in the 1.8 million individuals living in the Denver area. This number eventually found its way into newspaper reports as a risk of one in 900 million per individual. To reflect the likelihood of individual events as well as total consequences, the risk spectrum in Figure 5.4 compares the risk of Rocky Flats with that (as calculated in WASH-1400) of a single nuclear reactor and with the total of man-caused disasters in the United States.

Table 5.4

SUMMARY OF POSTULATED HIGHER RISK ACCIDENT SCENARIOS

EVENT	CMEC*	PROBABILITY (per year)	RISK
Building 559			
Structural damage resulting from an earthquake	5.6×10^{-3}	2.2×10^{-3}	1.2×10^{-5}
Tornado-driven automobile missile	48	2.5×10^{-7}	1.2×10^{-5}
Building 707			
Structural damage resulting from an earthquake	3.4×10^{-6}	5×10^{-2}	1.7×10^{-7}
Tornado-driven missile damage resulting in a			
plutonium oxide release from Module J	16.2	3×10^{-7}	4.9×10^{-6}
Structural damage caused by a tornado	25.1	5×10^{-7}	1.3×10^{-5}
Structural damage caused by an extreme wind	7.1×10^{-1}	1×10^{-3}	7.1×10^{-4}
A metal criticality in a glove box	1.4×10^{-2}	3.5×10^{-3}	5×10^{-5}
Building 774			
Tornado-driven missile damage	28	5×10^{-7}	1.4×10^{-5}
Ruptured pipe carrying low-level plutonium waste	7.9×10^{-3}	1.5×10^{-3}	1.2×10^{-5}
Building 776			
Structural damage resulting from an earthquake	2.6×10^{-3}	3×10^{-2}	7.8×10^{-5}
Structural damage caused by an extreme wind	2.3×10^{-1}	1.8×10^{-3}	4.1×10^{-4}
Wind-driven missile damage	1.0×10^{-2}	1.8×10^{-4}	1.8×10^{-6}
Building 779			
Structural damage resulting from an earthquake	6.6×10^{-4}	3×10^{-2}	2.0×10^{-5}
Structural damage caused by an extreme wind	1.0×10^{-1}	5×10^{-3}	5.0×10^{-4}
Wind driven missile damage	4.0×10^{-2}	3×10^{-4}	1.2×10^{-5}
Structural damage resulting from a tornado	1.66×10^{-1}	4×10^{-5}	6.6×10^{-6}
Plant Maximum Credible Accident			
Crash of a large fully fueled aircraft into a			
plutonium processing building	67	1.3×10^{-7}	8.7×10^{-6}
COMPOSITE RISK			2.0×10^{-3}

* Cancer mortality effects commitments

SOURCE: DOE (1983, 47).

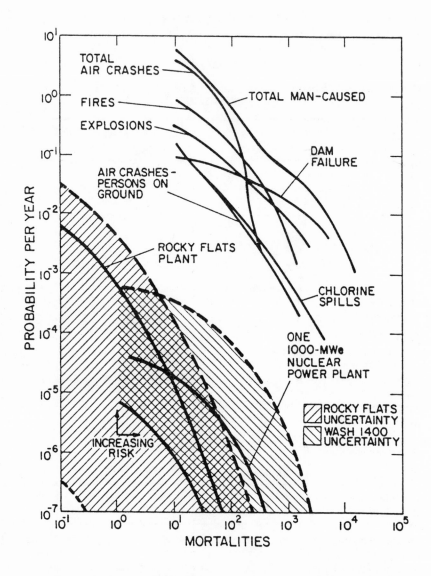

Figure 5.4 Comparison of rare event risks, including Rocky Flats, a nuclear power plant, and all man-caused disasters in the United States.

SOURCE: DOE (1983, 49).

A clear benefit of the risk analysis in the *Long-range Rocky Flats utilization study* is that it involved detailed estimates of the vulnerability of vital equipment. This leads to the conclusion that 97 percent of the risk assessed in the analysis of maximum credible and maximum probable events can be averted by engineering improvement costing $112 million over a 5-8 year period. Even if one doubts the accuracy of the estimate of total mortality risk, this finding offers a concrete way toward relative improvement and constitutes the most appropriate application of probabilistic risk assessment.

From a policy point of view, the comparison to reactor accidents is also informative. An examination of risk spectra for the two cases allows three interesting conclusions:

- For cancer as a consequence, the risk spectrum for Rocky Flats and the risk spectrum for a single water-moderated nuclear reactor are similar in shape.
- The maximum credible event for Rocky Flats (an airplane crash) appears, by a factor of 10, less damaging than the equivalent event for nuclear reactors.
- At the high-probability/low-consequence end of the spectrum, Rocky Flats substantially exceeds the risk spectrum for a nuclear power plant.

Over all, the probabilistic analysis of rare disasters at Rocky Flats is about as satisfying as the comparable analysis of nuclear power. It identifies important technical means for improving the technology and in this sense has direct utility. It also demonstrates that the worst scenarios that one can think of do not lead to large average risks—mainly because they are estimated to be highly unlikely. Unfortunately, in the wake of Chernobyl, the real achievements of disaster assessment at Rocky Flats do not quiet the fears of the critics any more than the *Reactor safety study* laid to rest fears about nuclear power.

Summary and Conclusions

Rocky Flats involves the manufacture of a product—nuclear weapons components—that is an extreme hazard. Production involves a materials process that itself poses the threat of extreme consequences. To manage and control both product and process, Rocky Flats hazard managers have designed a plant in which multiple barriers block the dispersal and theft of plutonium. To establish this strategy of defense in depth, they have allocated essentially unlimited resources to construct a facility in which nearly all investment is in some way related to hazard control.

Several conclusions relevant to overall industrial hazard management emerge from this study.

Defense in depth works. Where consistently used, such as in air filtration, multilayered defense in depth has succeeded in protecting workers and the public on a daily basis and has strongly limited damaging releases under accident conditions, such as the 1957 and 1969 fires. When not used, as in the storage of plutonium-bearing cutting-oil wastes, what occurred was probably the largest single plutonium release in the plant's 35-year history. If there is any general question about the strategy of defense in depth, it is whether ordinary, privately financed corporations can afford the scale of protection that prevails at Rocky Flats. Whereas the generic principle of defense in depth is transferrable, the scale of the multilayered Rocky Flats experience is probably not. Indeed, the Bhopal disaster (chapter 6) provides telling evidence that defense in depth can fail colossally.

Learning is important. Over the 35-year history of Rocky Flats, various plant managers have achieved notable improvements, including: a marked reduction in fire outbreak brought about by improvements in containment; a significant reduction in workers' radiation exposure, brought about by lower internal standards and redesign of operations; and a decrease in public exposure, assured by the enlargement of the public exclusion zone and improved filtration systems. The introduction of the new, $300-million plutonium building in 1982, brought further reductions in worker exposure and along with this, decreased the risk of public exposure. And improvements in buildings and machinery, identified through probabilistic risk analysis in the *Long-range Rocky Flats utilization study*, offer future opportunities for risk reduction in the case of catastrophic accidents brought about by natural events. At the same time, however, a recent appraisal of technical safety pinpoints a series of deficiencies (DOE 1988).

Public scrutiny can help. The studies of environmental impact and long-range utilization initiated through public concern have resulted in numerous useful findings. These include public and private measurements of plutonium in soil around Rocky Flats, Department of Energy analyses of possible occupational cancer, the critics' studies of possible plutonium-induced cancer in the public, and the Environmental Protection Agency's evaluation of plutonium burdens in the population around Rocky Flats. Taken together, these data give about as good an account of chronic hazard consequences as is available for any comparable hazardous industrial facility. In particular, it is apparent that occupational cancer among plutonium workers is not a major health problem and that the public cancer risk due to Rocky Flats plutonium is probably too small to measure.

Unanswerable questions remain. Despite enormous hazard management investments and substantial efforts in risk assessment, a number of important questions remain unanswered: (a) Can one rely upon the limited probabilistic risk analysis of the *Long-range Rocky Flats utilization study,* or does it contain errors of several orders of magnitude? (b) Are the risks of terrorism and plutonium theft small, or do they loom larger than the risks of ordinary operations? (c) Is the scientific consensus on the effects of internal alpha emmitters valid, or are we in for rude surprises as this complex subject continues to unfold? None of these questions will be easily answered, since each involves what Alvin Weinberg has characterized as "trans-scientific" issues (Weinberg 1972).

Nuclear weapons are nuclear weapons. Perhaps most important, Rocky Flats will not soon shake its image of a "death factory," at least for the sizable fraction of the population that fears and loathes nuclear weapons. One may safely predict that no matter how effective hazard management at Rocky Flats, the facility will contain to evoke intense distrust by individuals who have strong, emotional misgivings about the manufacture of nuclear weapons, about nuclear deterrence, and about the nuclear arms race. In particular, the demonstrators at the gate will not go away.

In one sense, Rocky Flats is an industrial hazard management success disguised as a failure. Plant management has over the years been able to hold occupational and public risk at a relatively low level, even while coping with large plutonium accidents inside the plant and carrying out a series of extremely hazardous activities. In achieving this success, plant management has been motivated by its own considerable safety concerns, a series of oversight commissions, several comprehensive and many narrowly focussed risk studies, and, most recently, a string of lawsuits by abutters. In response to DOE's recent concerns, Rockwell International has categorically stated that "safety *will not* be compromised for *any* reason" (DOE 1988, Appendix B, p. 2) and reaffirmed the company's goal of "making the Rocky Flats Plant a weapons complex leader in safety" (DOE 1988, Appendix B, p. 3).

It is hard to avoid the conclusion, therefore, that continuing controversy and misgivings about safety at Rocky Flats are linked more to the products of the plant than to the process by which they are made. As long as people carry with them the nightmare of nuclear war, it seems unlikely that the present negative image of Rocky Flats will change—no matter how successful its hazard-management programs.

6

Union Carbide Corporation
and the Bhopal Disaster

Runaway chemical reactions are rare events, particularly in this heyday of the redundant and "defense-in-depth" safety design for complex, high-risk technologies described so well in the last chapter. Yet during the chill of night between December 2 and 3, 1984, a statistically improbable worst-case scenario moved from the computer simulations of the risk assessors and played itself out on the unsuspecting citizens of Bhopal, India. A parade of corporate failures—in design, in maintenance, in operation, in emergency response, and in hazard management—conspired with a southerly wind and a temperature inversion to push a lethal cloud of methyl isocyanate (MIC) out to kill and injure thousands of people, animals, and plants in the area (Figure 6.1). By sunrise, the unprecedented horror had catapulted Bhopal to the head of history's roll of industrial disasters (Table 6.1) through 1984 (the 1986 Chernobyl accident may eventually eclipse the Bhopal toll).

Much is at stake in corporate responses to and learning from the accident, for the post-mortems may select the "wrong" lessons and thus fail to avert future calamities, place unwarranted crippling restraints on the chemical industry, or impede the flow of needed and generally beneficial technology to developing countries. The chemical industry, as noted in chapter 2, with a job-related total recordable incidence of 2.26 fatalities, injuries, and illnesses per 100 full-time workers in 1985 (compared with an all-industry incidence of 6.31), has been and still is an undisputed leader in industrial safety (National Safety Council 1987, 40). Union Carbide Corporation, the parent company involved in the disaster at Bhopal, has more than twenty years' experience in the safe manufacture, use, transport, and storage of MIC (to say nothing of a host of other hazardous products). With a cadre of scientists and technicians and an institutional structure for environmental protection, India is better equipped than other developing countries to manage hazardous technologies. Given this context, other industries in

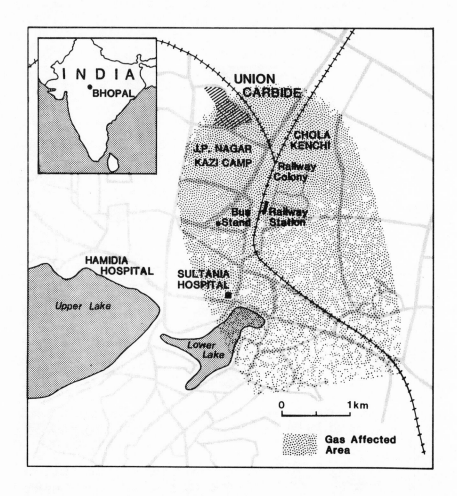

Figure 6.1   Extent of MIC dispersal at Union Carbide's Bhopal site.

SOURCE: Clark University Cartography Lab.

Table 6.1

SIGNIFICANT INDUSTRIAL DISASTERS
OF THE TWENTIETH CENTURY: A CHRONOLOGY

YEAR	LOCATION	INCIDENT	MAJOR CONSEQUENCES
1921	Oppau, Germany	explosion in chemical plant	561 deaths
1942	Honkeiko, China	coal-dust explosion	1,572 deaths
1947	Texas City, Texas, USA	fertilizer-ship explosion	562 deaths
1956	Cali, Colombia	dynamite-truck explosion	1,100 deaths
1959	Minamata, Japan	discharge of mercury into waterways	400 deaths; 2,000 injuries
1973	Fort Wayne, Indiana, USA	rail accident involving vinyl chloride	4,500 evacuated
1974	Flixborough, UK	explosion in a chemical plant	23 deaths; 104 injuries; 3,000 evacuated
1975	Chasnala, India	mine explosion	431 deaths
1976	Seveso, Italy	dioxin leak	193 injuries; 730 evacuated; hundreds of animal deaths, 200 cases of skin disease
1978	Los Altaques, Spain; Xilatopec, Mexico; Manfredonia, Italy	propylene spill in truck accident; gas explosion in road accident; release of ammonia	216 deaths; 200 injuries; 100 deaths; 50 injuries; 10,000 evacuated
1979	Three Mile Island, Pennsylvania USA; Novosibirsk, USSR; Mississauga, Ontario, Canada	nuclear reactor accident; chemical-plant accident; release of chlorine and butane in rail accident	200,000 evacuated; 300 deaths; 200,000 evacuated
1980	Norway	collapse of off-shore oil rig	123 deaths
1981	Tacoa, Venezuela	oil explosion	145 deaths; 1,000 evacuated
1982	Taft, Louisiana, USA	explosion involving acrolein	17,000 evacuated
1984	Cubatão, São Paulo, Brazil; San Juan Ixhuatepec, Mexico; Bhopal, India	pipeline explosion, release of petrol; natural-gas explosion; leak of poison gas	508 deaths; 452 deaths; 4,248 injuries; 300,000 evacuated; 2,998->4,700 deaths; 50,000 severe injuries; 200,000 evacuated
1986	Chernobyl, USSR; Basel, Switzerland	nuclear reactor accident; fire of pesticide plant	>28-28,000 deaths; serious pollution of Rhine River
1987	Kotka, Finland	spill of monochlorobenzene in harbor	pollution of sea floor
1988	Aberdeen, Scotland	fire/expolsion on oil rig	>166 deaths, >71 injuries

SOURCES: Lagadec (1982, 503-512), *Financial Times* (1984, 1), Kottary (1985, 8), Shrivastava (1987, 12-13), Hohenemser and Renn (1988, 44), and UNEP (1988).

other places are more likely candidates for catastrophic disasters. Thus it is essential to understand how and why this particular surprise occurred at Bhopal if we are to ward off future similar tragedies.

This chapter taps findings in an earlier study (Bowonder, Kasperson, and Kasperson 1985) and inquires into the causes of the Bhopal accident and its lessons for the corporate management of health and safety hazards. A profile of Union Carbide Corporation and its hazard-management programs, as they existed at the time of the Bhopal accident, sets the context. An examination of the decision to locate the Union Carbide plant at Bhopal leads to an analysis of precursors to the accident, the accident itself, and its consequences. The chapter closes with an assessment of the implications of the accident for corporate hazard management.

Union Carbide Corporation and Its Hazard-Management Programs

In 1984 Union Carbide Corporation, with total sales of $9.5 billion and 98,366 employees worldwide (Union Carbide Corporation 1985), was one of the largest chemical companies in the United States. It was also a diversified company, with 28 percent of its sales in petrochemicals, 26 percent in technology, services and specialty products, 20 percent in consumer products, 16 percent in industrial gases, and 10 percent in metals and carbon products. Its activities were spread over some 460 domestic and 270 foreign locations, in 38 different countries. The scale of the international operations is suggested by the fact that approximately one-half of all employees and nearly one-third of all sales came from the international segment of the corporation.

The several years preceding the Bhopal accident were not good ones for the chemical industry. But 1984 had been a reasonably good year for Union Carbide. Despite a slowdown in the second half of the year, annual sales were up 6 percent over the previous year, as net income increased from $79 million to $323 million. Warren Anderson, Chairman of the Board, could point to expanding activities in China, commercialization of 12 new high-performance specialty products, completion of the U.S.'s largest polysilicon plant, and further progress in the corporate plan to increase its reliance on technology-intensive businesses (Union Carbide Corporation 1985).

Union Carbide's hazard-management structure and programs were quite characteristic of those prevailing among large companies in the U.S. chemical industry. The top management official directly responsible for health and safety reported to a senior vice president. It is worth noting the contrast with the situation at PETROCHEM, where the chief official for health, safety, and environment reports directly to the President (see chapter 2). As at PETROCHEM, hazard management at Union Carbide had undergone substantial corporate upgrading in the mid-1970s, when various health, safety, medical, and legal functions were centralized into a single management unit. In 1984, the health and safety and environmental affairs unit had 130 staff in corporate headquarters and 700 total in the corporation. The unit had four subdivisions—environment (with access to various computerized data bases), health and safety, medical (including several epidemiologists and biostatisticians), and product safety (including three toxicologists but with access also to the Bushe Run toxicology plant, with its staff of 80 people). The corporate hazard-management structure also involved a company doctor at all plants and an industrial hygienist at all major plants.

In terms of its corporate regulatory system, the company shows parallels with the situation at PETROCHEM. Union Carbide accepted all external regulatory standards and supplemented them with its own internal corporate standards, which were the primary responsibility of a health-effects review group. Each of the corporate standards, which management officials numbered at close to 100, was supported by a "rationale document" that was submitted to broad peer review in the corporation. In addition, each plant had its own annual safety goal, a hazard profile, and a hazard profile rating (at one of three levels). Unfavorable trends in progress toward the goal or in the safety profile could trigger a health and safety audit of the plant.

The audit system differed in structure and implementation from that at PETROCHEM (chapter 2). Union Carbide Corporation in 1984-85 conducted three types of audits: product-safety audits, compliance audits (at plant locations), and chemical process audits. It took particular pride in its chemical process audits, for which it viewed the corporation as an industry leader. Plant-level audits depended upon the hazard rating. In 1985, the corporation had 58 locations in its high-hazard rating (level 1), 52 in its medium-hazard rating (level 2), and 620 in its low-hazard rating (level 3). High-hazard locations would be audited by a 4-5 member team, typically for a 10-day period; medium-hazard locations by a 3-5 member team for 6 days, and low-hazard locations by a 1-3 member team for 4 days. Characteristically, high-hazard locations were audited about every two years. A formal audit report detailing hazard issues was provided upon exit and had to be responded to by the location management within 60 days.

In short, Union Carbide had in place a substantial hazard-management capability, as well as organizational structures and processes, that had achieved an effective occupational safety record and that would appear equal to the task of preventing the disaster that occurred at Bhopal. After examining the circumstances under which Union Carbide came to Bhopal, the authors take up this question.

Union Carbide Comes to Bhopal

Union Carbide was scarcely an unwelcome intruder in Bhopal. The Indian government promoted the siting of industries in less-developed states, such as Madhya Pradesh, where Bhopal is located. Madhya Pradesh leaders offered incentives to companies that would bring jobs and indigenous manufacturing to its unindustrialized cities; Union Carbide, for example, built on government land for an annual rent of less than $40 an acre. A plant that would manufacture the carbaryl pesticides to fuel India's ongoing green revolution was particularly welcome as another step toward self-sufficient food production. Hence the 1970 decision of Union Carbide of India Limited (UCIL) to manufacture the pesticide Sevin in an advanced facility in central India met with great fanfare.

Sevin, manufactured from MIC, had received the endorsement of the Indian Council of Agricultural Research. Use of the pesticide decreases insect damage of cotton, lentils, and other vegetables by as much as 50 percent. Even in the several years since the accident, few serious observers have suggested that India do away with Sevin and other carbaryl pesticides, which, ironically, are substitutes for "more dangerous" DDT and organophosphates. Given the high toxicity of MIC (see Table 6.2) however, it is clear that the chemical requires, at all stages, special handling commensurate to the risk.

It is easy to contend that high-risk facilities have no place in densely populated urban areas. Yet such a facility—be it a liquefied-natural-gas facility in Mexico City, a petrochemical complex in Cubatão, Brazil, or a pesticides factory in Bhopal—is apt to attract squatter settlements to its gates. The showpiece UCIL factory and other industries that set up shop in Bhopal surely contributed to the staggering rise in population—from 350,000 in 1969, to 700,000 in 1981, to over 800,000 in 1984.

The showplace factory never lived up to its promise of production and jobs for the area. A drought in 1977 forced many farmers to take out government loans, many of which began to fall due in 1980. The farmers then exchanged the expensive Union Carbide pesticides for other less costly and less effective alternatives. Meanwhile, the Indian government was providing incentives for small-scale manufacturers to produce pesticides that they could afford to sell at half the

Table 6.2

WORKPLACE LIMITS TO CHEMICAL EXPOSURES

CHEMICAL	PARTS PER MILLION[a]
Carbon monoxide	50.00
Chloroform	25.00
Methylamine	10.00
Benzene	10.00
Acetic acid	10.00
Cyanogen	10.00
Phosgene	0.10
Methyl isocyanate	0.02

[a] Time-weighted averages for 8-hour exposure

SOURCE: Data from ACGIH (1984).

price of Union Carbide products. In addition, inexpensive, nontoxic synthetic pyrethroids made their debut, and sales of traditional pesticides began to drop throughout the industry. The Bhopal operation, never very profitable, broke even in 1981 but thereafter began to lose money. By 1984 the plant produced less than 1,000 of a projected 5,000 tons, and lost close to $4 million. UCIL, contemplating selling the operation, began to issue incentives for early retirement and to cut back on its workforce. Many of the skilled workers moved on to securer positions. Things were not going well.

Early Warnings

Whether cost-cutting measures and the departure of skilled personnel caused lapses in safety is difficult to ascertain. Nevertheless, the Bhopal plant experienced six accidents—at least three of which involved the release of MIC or phosgene, another poisonous gas—between 1981 and 1984. These accidents scarcely presaged the catastrophic release, but, taken together, they surely pointed to safety problems at the plant. Indeed, a phosgene leak that killed one worker on December 26, 1981 generated an official inquiry, but the findings (filed three years later) gathered dust in the Madhya Pradesh labor department until after the Bhopal accident, when two officials lost their jobs for having failed to act upon the report's safety recommendations.

Meanwhile, a local journalist warned that the plant's proximity to Bhopal's most densely populated areas was inviting disaster. In 1982, Rajkumar Keswani took on UCIL in a series of articles in the Hindi press, "Sage, please save this city," "Bhopal on the mouth of a volcano," and "If you don't understand, you will be wiped out," the headlines warned (Keswani 1982 a,b, c). On June 16, 1984, he tried again, this time with what he calls "an exhaustive report on the Union Carbide threat." "The alarm fell on deaf ears," he wrote one week after the Bhopal accident (Keswani 1984).

A 1982 safety audit by an inspection team from Union Carbide cited a number of safety problems, including the danger posed by a manual control on the MIC feed tank, the unreliability of certain gauges and valves, and insufficient training of operators (Kail, Poulson, and Tyson 1982). UCIL claims to have corrected the deficiencies, but auditors have never confirmed the corrections.

Just before the accident in Bhopal, Union Carbide Corporation's safety and health survey of its MIC Unit II plant in Institute, West Virginia, cited 34 less serious and two major concerns, the first of which was the "potential for runaway reaction in unit storage tanks due to a combination of contamination possibilities and reduced surveillance during block operation" (Union Carbide Corporation Engineering and Technology Services 1984). Why the parent company, which owned 50.9 percent of the Bhopal plant, failed to share with its subsidiary its two major concerns (the second was the serious potential for overexposure to chloroform) is unclear. Some Union Carbide officials contend that the different cooling systems—brine at Institute and freon at Bhopal—made the hazard communication unnecessary, but this is difficult to square with the recommendation:

The fact that past incidences of water contamination may be warnings, rather than examples of successfully dealing with problems, should be emphasized to all operating personnel. (Union Carbide Corporation Engineering and Technology Services 1984)

Equally puzzling is the parent company's earlier overriding of an alleged UCIL protest against the installation of such large storage tanks—15,000 gallons—at Bhopal (Brushnan and Subramanian 1985, 109).

In any event, MIC sat in storage at the Bhopal plant for at least three months prior to the accident. Such storage courts calamity, for the tiniest ingress of water, caustic soda, or even MIC itself is sufficient to set in motion an exothermic (heat-producing) chemical reaction (Worthy 1985, 28).

The Accident

Some time shortly before th 10:45 p.m. shift change at the Bhopal plant on December 2, 1984, water, another contaminant, or both, entered MIC storage tank 610 (see Figure 6.2) thereby triggering a violent chemical reaction and a dramatic rise in temperature and pressure. It is not known whether the incoming control room operator was aware that the 10:20 p.m. tank pressure read 2 psi (pounds per square inch), but the 11:00 p.m. reading of 10 psi does not seem to have struck anyone as unusual. Nor should it have, since normal operations ran at pressures between 2 and 25 psi.

By the time the operator did take notice of the rising pressure—from 10 psi at 11:00 p.m. to 30 psi at 12:15 a.m.—the reading was racing to the top of the scale (55 psi). Escaping MIC vapor ruptured a safety disc and popped the safety valve. On the heels of this initial release came a series of compromises and failures of virtually all the safety systems designed to prevent release. The deadly gas spewed out over the slums of Bhopal.

Some of the details surrounding the release remain sketchy, yet it is possible to construct a reasonably plausible analysis of the accident. Figure 6.3 depicts the accident in terms of the causal model of hazard described in chapter 1.

In the case of the Bhopal accident, the basic *human need* for food generated a *human want* of increased food supply through pest control (i.e., the manufacture and application of pesticides). The particular *choice of technology* at Bhopal was the indigenous manufacture, using MIC, of carbamate pesticides. As Figure 6.3 indicates, this choice of technology entailed, at least implicitly, a series of important choices, ranging from the basic selection of chemical over biological pest control to a series of considerations relevant to the storage of toxic materials and the scale of technology. These decisions fundamentally shaped the inherent hazard that was set off by the *initiating event*, the contamination of MIC.

All the while, standard means to *prevent exposure*—remote siting, exclusion zones or so-called greenbelts, early-warning and emergency-response systems, evacuation plans, and hazard communication—never materialized at Bhopal. The surprise release of poison gas thwarted any concerted effort to *prevent consequences*. For want of instructions to breathe through a wet towel, scores of people died ghastly deaths. For lack of information on the toxicity of MIC,

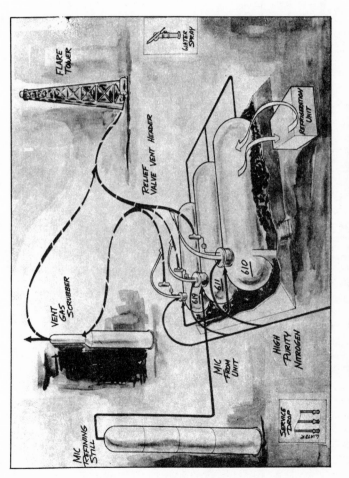

Figure 6.2 MIC tanks at the Bhopal pesticide plant.

SOURCE: Union Carbide Corporation.

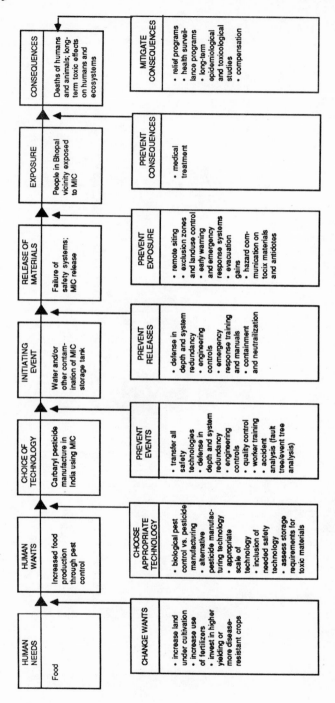

Figure 6.3 The causal structure of hazard: Application to the Bhopal accident.

Bhopal's medical community was hard put to *mitigate consequences*. Figure 6.4 shows how these contributors at each stage of the accident in effect hurried a low-probability hazard along to catastrophic consequences.

Within hours, 200,000 people of Bhopal were exposed to MIC, with 50,000-60,000 receiving substantial exposure. Although the full death toll will never be known, the official government estimates of about 1800 undoubtedly underestimates substantially the number of fatalities (Bowonder, Kasperson, and Kasperson 1985; Lepkowski 1985). Four years after the accident, the chronic effects of the accident continue to exact their tolls on the heavily exposed people of Bhopal and on the local ecosystem (*Washington Post* 1988, A26).

Corporate Hazard Management

The Bhopal accident raises a number of basic issues for how corporations manage industrial hazards. Some of these have elicited substantial responses since Bhopal; others have not. In the discussion to follow, the authors take up some of the more prominent issues.

Choice of technology. Few basic decisions affect hazard potential more than the initial choice of technology. In the case of Bhopal, the choice involves the long-term storage of MIC, a chemical so extremely hazardous that some countries expressly prohibit long-term storage. Bayer AG, a large multi-national corporation, manufactures MIC in West Germany and in Belgium, but the process uses the nontoxic intermediates dimethyl urea and diphenyl carbonate and involves no dangerous phosgene or chlorine. Moreover, Bayer promptly converts MIC into end products that are safe to store (Worthy 1985, 32), and temperature and pressure gauges on the tanks automatically control inconsistencies and can immediately trigger an alarm system (Ramaseshan 1984). France prohibits domestic manufacture of MIC and requires that special MIC storage drums be maintained in separate sheds equipped with automatic water sprinklers and sensitive gas detectors (Delhi Science Forum 1985, 46). England allows only one company to handle MIC and the gas must be stored at a site two miles out of the town of Grimsby (Delhi Science Forum 1985, 46). Such alternative technologies, replete with added automated safeguards, pose a lower inherent risk of catastrophic releases than the dangerous process chosen for the Bhopal plant.

Until 1978, UCIL did not store MIC at Bhopal. At that time, U.S. corporate headquarters decided, apparently to promote efficiency, to utilize a technology that favored large inventories of MIC, despite its high toxicity and in the face of reservations from the Indian subsidiary. The decision to store MIC at Bhopal was taken without apparent considerations of the particular safety issues posed by a location in India as compared with one in North America. In addition, the Bhopal plant relied on manual, labor-intensive controls whereas Union Carbide's plant in Institute, West Virginia, used a computerized monitoring system. These basic choices about the technology to be employed and the extent of needed safeguards at the Bhopal plant contributed ultimately to the disaster that occurred.

In the several years since Bhopal, corporations have learned some generic lessons. Nearly all the large chemical companies in the United States and

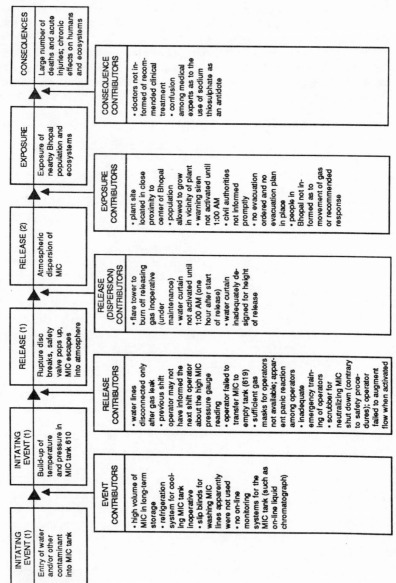

Figure 6.4 Detailed model of contributors to the Bhopal accident.

Europe have reexamined safety practices at their plants. Many have discovered problems with maintaining large stores of hazardous chemicals and accordingly reduced inventories at key hazard points. Whether or not the broader issues involved in hazard control by technology choice have been confronted is more uncertain.

The choice-of-technology mistake at Bhopal by Union Carbide was not the transfer of the modern formulation plant for bulk pesticides needed for Indian agriculture. Rather it was the construction of such a large MIC-based plant of this particular design (the most advanced pesticide plant in the developing world); the choice of a particular production process that carried higher inherent risk for a catastrophic accident; the choice of a scale of plant and equipment (designed to produce 5,000 tons a year) that involved large inventories of highly toxic substances; and the failure to incorporate state-of-the-art technological designs and practices for defense-in-depth protection against major accidents.

In the future, near-term economic considerations alone should not drive corporate choice of the technology. Such considerations should be meticulously balanced with opportunities for risk minimization and long-term corporate protection.

Siting hazardous industrial facilities. Bhopal sits in the middle of central India, and the plant site is astride main rail lines leading in all directions. From Union Carbide's perspective, the good location also came with electricity, an ample water supply for manufacturing chemicals, and a ready pool of labor.

Since India had no policy on the siting of hazardous industries, no one contested the location of the plant a scant two miles from the center of the city (see Figure 6.1). Although the state department of Town and County Planning was aware of the plan to manufacture and store MIC, it approved the location and, in preparing the Bhopal Master Plan in 1975, subsequently classified the plant as "general" rather than "obnoxious" and did not require it to relocate to a more remote site (although 16 other smaller industries were relocated (Qureshy 1985). Indeed, in the aftermath of the accident, the Indian press engaged in controversy over whether an official who may or may not have issued an eviction order in 1975 was consequently transferred (Qureshy 1985).

The absence of an exclusion zone or land-use control around the plant exacerbated the high risk of the location in one of the fastest-growing cities in the world. Between 1961 and 1981 Bhopal's population increased by nearly 75 percent per decade, crowding people living in primitive huts onto government lands. The state government, having failed to prevent the population growth adjacent to the plant, compounded the problem by conferring legal status on those living in the squatter settlements in April 1984 so that they could vote in the December elections. This action was taken with knowledge of the accidents that had occurred at the plant and of the highly critical results of the 1982 safety investigation conducted by Madhya Pradesh.

Past industrial disasters, including the Mexico City explosion, the Three Mile Island accident, and Chernobyl, have taught that remote siting provides a major protection against potential industrial catastrophes. Particularly where engineered safeguards may be compromised through poor maintenance and equipment failure or operator error, physical distance—by affording greater dilution of the release and increased time for responses—offers an overall redundancy in safety. The tragic loss of life at Bhopal was concentrated in the densely populated squatter settlements located within three kilometers of the plant. A remote site for the plant or an exclusion zone of three kilometers at the Union Carbide plant might have averted most of the fatalities.

One apparent message from Bhopal would call for corporations to practice remote siting of facilities that carry the potential for rare but catastrophic events. Such an approach will need cooperation from the public sector to control population growth around the facility. Short of truly remote sites, well-considered locations and land-use controls can provide important means of disaster minimization. Exclusion zones such as the 11 square miles surrounding Rocky Flats (chapter 5) similar to those surrounding nuclear plants (in India as well as elsewhere) should be provided, with the size of the zone determined by comprehensive, site-specific risk analyses.

Defense in depth. The Bhopal plant employed defense in depth, which, although not as extensive as that employed at Rocky Flats (see chapter 5), involved layers of safety systems intended to prevent major releases even in the face of individual system failures. Its five major safety systems were:

- **the refrigeration system.** The MIC storage tanks were connected to a refrigeration system that circulates the liquid MIC and keeps it cool. In the event MIC becomes contaminated, the refrigeration slows the reaction that may occur, thereby increasing the time available for safety response.
- **the spare tank.** One of the three 60-ton tanks at the plant is always left empty so that, in the event of an accident, MIC from a leaking tank can be diverted to the spare.
- **the flare tower.** A 30-meter-high pipe located a short distance from the MIC unit is used to burn toxic gases high in the air, thereby rendering them harmless.
- **the vent gas scrubber.** A tall, rocket-shaped unit is intended to detoxify any releasing gas by spraying it with caustic soda solution and converting it into a harmless vapor.
- **the water curtain.** The plant was equipped with a network of waterspouts that, in the event of an accident, shoots jets of water 12 to 15 meters in the air forming a water curtain around the gas leak. The water neutralizes the MIC vapor to form dimethyl urea or trimethylbiuret, both comparatively safe substances.

Even if one or several of these systems were to fail, the others should successfully protect against any massive offsite release.

The Bhopal accident testifies to the vulnerability of even a well-founded corporate safety philosophy to diverse implementation failures and human error. At the time of the accident, with or without Union Carbide's authorization, the refrigeration system was not working. The use of a requisite spare tank—which may or may not have been empty—required that the operator manually open the valves connecting the two tanks, an operation taking no more than three minutes. In the confusion of the accident, however, the valves were not opened. The vent gas scrubber, on standby mode since October 23, 1984, failed, possibly because the operators neglected to augment the flow of caustic soda required to neutralize MIC. The flare tower designed to burn off escaping gas was under maintenance (because of pipe corrosion) and thus inoperative. And the spouts designed to shoot jets of water into the air to quench a gas leak could not cope with the gusher of MIC some 35 meters high. In short, design errors, sloppy maintenance, poor safety practices, inadequate operator training, and human error hopelessly compromised a many-layered safety system that should have worked at Bhopal, as it works at Rocky Flats (chapter 5).

Less than a year after the Bhopal calamity, a leak at another Union Carbide plant dampened the "can't-happen-here" claims of the U.S. chemical industry. The leak on August 11, 1985, of methylene chloride and aldicarb oxime at Union Carbide's plant in Institute, West Virginia, attests to the vulnerability of back-up safety systems that fail even in the face of $5-million worth of upgrading. No technological or managerial quick fix is equal to overcoming the vulnerability, pervasive even in well-conceived safety systems; the flaws in design; the inadequacies in computer modeling (or failure to input relevant data); the human reluctance to acknowledge failure when it does occur (no one wants to sound the siren); and the intrinsic limitations in the depth of regulatory inspection, review, and enforcement.

Emergency preparedness. Bhopal was ill-prepared for the disaster. No emergency manuals or evacuation plans were available to local officials who, along with nearby hospital officials, were not aware of the toxic substances at the plant, their degree of toxicity, potential health effects, or the recommended medical treatment. The exposure had occurred before the warning siren sounded (some hours after the beginning of the release), and no public warnings were issued. No information on the movement of the gas cloud—broadcast to the factory workers—alerted people in the vicinity about the best direction in which to flee. Tragically, people poured out of their homes and many ran toward the factory. Although Union Carbide Corporation headquarters sent a telex on December 5, 1984, indicating that victims might be given amyl nitrite, sodium nitrite, or sodium thiosulphate if cyanide poisoning (a possible sequela to MIC exposure) were suspected, UCIL did not divulge it to the public, arguing that administration of an antidote for cyanide would create widespread panic (Ramaseshan 1985). Even basic instructions to breathe through a wet towel, which could have saved the lives of hundreds, were never forthcoming.

Hazard communication and response. Some of the clearest failures in the Bhopal accident involved inadequate information flow and emergency preparedness. These failures occurred at various levels: inadequate worker understanding of MIC's toxicity and health threat; lack of knowledge by local government and medical officials of the plant's chemicals and their hazards; poor information during the accident to guide nearby residents; and lack of advice to local medical personnel as to recommended treatments. There was also no evacuation plan, and some workers may have escaped danger only because the door they chose to flee through happened, fortuitously, to open north rather than south.

The accident at Bhopal drove home an important lesson about the critical importance of effective hazard communication programs and about the formidable social, cultural, and institutional impediments that beset them, particularly in a developing country. Such roadblocks include slow bureaucratic response, cultural differences in experience with chemicals, and underdeveloped social communication networks. That the state government convened only at its normal meeting time, ten hours after the accident, to handle the emergency indicates the need to change the basic societal structure for responding to hazards. Bhopal had only one telephone per one thousand people, running water for only a few hours per day, few street signs or traffic lights, and crowded 12-foot-wide thoroughfares in which cows, goats, water buffaloes, taxis, and horse-drawn carriages moved simultaneously in both directions (Diamond 1985, 8). Local residents, with little experience with chemicals as part of everyday life and with no direct information provided by the plant, viewed the plant as producing "medicine for the crops" and not substances harmful to people.

As a direct result of the Bhopal accident, the Chemical Manufacturers Association (CMA) initiated its Community Awareness and Emergency Response (CAER) program. Launched in March 1985, this program called upon member companies to develop, with extensive community involvement, emergency plans for individual chemical plants and to test them. By the end of 1986, some 170 member companies were implementing CAER programs at 1500 sites; 600 sites had completed or were completing emergency plans; more than 230 CAER plans had been approved by local officials; and 130 communities had tested their plans (CMA 1987; Cathcart 1987). Despite evident variability in the quality and effort given by individual companies, the program had registered sufficient success to win the endorsements of the National Safety Council and the Red Cross (Cathcart 1987).

Risk assessment. The past ten years have witnesses a major development in the methods of probabilistic risk analysis (PRA), particularly for the nuclear industry (Kasperson *et al.* 1987). Other industries, including the chemical industry, have been slow to embrace these new analytic methods. The U.S. Nuclear Regulatory Commission has summarized the status of these techniques and their potential contributions for plant design and maintenance (NRC 1984). These methods are likely to be more extensively used in Europe as chemical plants implement in 1988 the various safety studies required by the Seveso Directive, the European Economic Community's response to the 1976 accidental dioxin release at Seveso, Italy. And the recent requirements for hazard communication and emergency planning in the 1986 amendments to Superfund (SARA) may lead to greater use of PRA by American industry.

Although the absolute numerical values produced by such analyses should be viewed cautiously, the PRAs now being widely performed for nuclear power plants (and some other industrial facilities) have considerable power for identifying potential accident sequences that could lead to catastrophic consequences. These analyses should be more widely used in the licensing process for technologies with catastrophic-risk potential. They should be comprehensive in scope—covering manufacturing operations, intermediate steps, storage, and transportation of materials—and should more effectively integrate behavioral and cultural considerations, going beyond current practice. (Even in the nuclear case, human behavior data bases remain underdeveloped).

A comprehensive risk analysis, such as that for the Canvey Island petrochemical complex in England (U.K. Health and Safety Executive 1978, 1981) or the Rijnmond petrochemical authority in Holland (Rijnmond Public Authority 1982) should not only assess major risks but also identify cost-effective opportunities for risk reduction. The chemical industry will need in the coming years to upgrade its practices and capabilities in risk assessment; the recently founded Center for Chemical Process Safety and the Chemical Industry Institute of Toxicology program in risk assessment (Starr 1987) are steps in the right direction. The former's recently published *Guidelines for hazard evaluation procedures* (AIChE 1987) suggests one useful role for industry-wide initiatives.

Concluding Note

The accident at Bhopal vividly points to needed improvements in corporate hazard-management programs. Multinational corporations need to do more than conduct the initial reviews of toxic substance storage and handling as they did in the months following the accident. They need to reexamine, and undoubtedly increase, their capabilities in formal risk assessment for the safe siting, design, and management of plants. Most will require new resources to appraise both catastrophic risk and the relevant community settings at plants. This upgrading could well be modeled after the Institute for Nuclear Power Operations, which was established after the Three Mile Island incident, and should consider the use of simulators in operator training. The corporate codes of social responsibility will also need to address explicitly the long-term lessons from Bhopal and act on the more demanding obligations by substantially improving auditing, monitoring, and compliance programs.

7

Corporate Management
of Health and Safety Hazards:
Current Practice
and Needed Research

Corporate management of health and safety hazards, chapter 1 argues, is terra incognita. As with all unknown lands, myths and stereotypes abound, shaped in this case by the extensive chronicling of management failures, vivid risk events, and images of corporate behavior. The preceding chapters do not provide a road map to the varied topography of widely differing industrial sectors, long established and new technologies, large and small firms, profitable and marginal companies. Rather, the case studies seek to detail and compare the characteristics of hazard management programs in five large corporations in the 1980s. The corporations tend to lie at one end of the industrial continuum in terms of size and resources available for hazard management. Several are unquestionably among the leaders of industrial health and safety; one experienced one of the major failures of the twentieth century. So the landscape of industrial hazard management depicted here is complex, featuring both industrial success and failure, opportunity and limitation, and emerging trends in safety as well as past errors

This concluding chapter begins by drawing together from the case studies the generic empirical characteristics of current practice in corporate management of hazards. Of course, much more needs to be known; accordingly the second portion of the chapter offers an agenda of research needs.

Characteristics of Current Practice

A larger hazard burden. Prior to the 1970s, managing technological hazards was never a major undertaking within industry. With a society much less environmentally and health conscious, with scientific knowledge of hazards underdeveloped, and with regulatory structures embryonic, employers were primarily geared to the obligations entailed by common law—to communicate to the employee the hazards involved in a work activity and to provide to employees and publics reasonable protection from harm. Acute hazards defined much of what constituted an employer's knowledge of dangers to employees and to the environment. And all too frequently in the marketplace the adage was "let the buyer beware!"

Not so with the today's large corporation. A single modern corporation confronts an oft-bewildering multitude of hazards—to employees, to consumers, to plant neighbors, to the environment—which it must assess and manage. This is no longer a side activity to be run out of the plant manager's back pocket or a minor office in the corporate division of public affairs. PHARMACHEM (chapter 3), for example, combines the occupational hazards of a small-parts assembly plant with the complex hazards involved in chemical storage, synthesis, and blending. The production of one PHARMACHEM product involves some 30 chemicals of varying hazard and poses, in a single product, a substantial burden in assessing and managing hazards, including potential accidental release, combustion, absorption, inhalation, or ingestion.

Meanwhile, PETROCHEM (chapter 2) must deal not only with the long-standing risks of accidents at drilling rigs and with fires and explosions involving petroleum products but with the chronic hazards associated with the 1,500 chemicals involved in its chemical divisions and newer less-understood chronic hazards of inhalation of gasoline vapors. Although Volvo (chapter 4) in its early history could concentrate on the hazards posed by automobile collisions and industrial accidents, it must now deal with the more perplexing issues involved with automobile emissions, the use of new materials in the manufacturing of cars and trucks, and increasing automation. The industrial process at Rocky Flats (chapter 5) uses basic material so toxic that hazard management consumes most of the company's total effort. Union Carbide found at Bhopal (chapter 6) that even an extensive hazard-management program proved inadequate to prevent disaster with a toxic substance such as methyl isocyanate.

Yet whereas hazard management commands increasing resources and a complex organization for many corporations, for many others in industry the picture is quite different. For them—as for the workers at the local gas station, for migrant pickers, for meatpackers, or for construction workers—hazard management is still a peripheral and neglected activity. Probably all industries have experienced some upgrading of hazard responsibility over the last several decades, but for many of the large manufacturing corporations the changes have been dramatic. Here hazard management has become a major corporate activity, replete with a degree of professionalization, specialization, and demand for expertise formerly confined to the governments of nations and states.

A growing recognition. As the burden of hazard management has escalated, so too has the view of safety, waste reduction, and environmental protection. Society's values about environmental protection and health security have changed dramatically over the past several decades. The complex institutional structure of regulatory agencies and environmental groups reflects these changes. It is not surprising that, whether by persuasion or necessity, corporations have evolved a new commitment to health and safety. Since the 1970s, corporate statements and codes of social responsibility have become commonplace in industry. The Harvard Business School has accorded the study of ethical issues a place in the training of business executives. Cradle-to-grave programs of protection have emerged in the chemical industry. Product stewardship and protection of workers have become more accountable and business managers have increasingly become liable, under both civil and criminal law, for the way they exercise these responsibilities. The Occupational Safety and Health Administration (OSHA) recently drove this point home to the Chrysler Corporation by fining the company $1.5 million for a series of job-safety violations (Holusha 1987).

It is abundantly clear from the continuing presence of industrial hazards and notable cases of management failure that serious inadequacies persist It is, for example, very uncertain what effect (if any) codes of social responsibility have had on corporate attitudes to hazard management. But it is also the case that the internal corporate expectation of its responsibilities is not the same in 1988 as it was in 1958 or 1968. Though stated goals generally preceded actual changes in behavior, and, of course, some firms are innovators in safety, others followers, and other highly resistant to change, a growing corporate recognition of its responsibilities for safety and environmental protection across the board is one of the realities of current trends in industrial hazard management.

A revolution in resources. It is scarcely surprising that industry possesses substantial resources for hazard management and that a major development in capabilities has occurred over the past decade, the preceding chapters indicate (1) how extensive this has been within some sectors of industry (particularly the nuclear power and chemical industries), but also (2) how variable it is within and among industries.

Since the case studies in the previous chapters involve large and wealthy firms, they tend to highlight the more dramatic examples of change and capability. This is most apparent at PETROCHEM (chapter 2) where the past 15 years have seen the health-and-safety staff at corporate headquarters grow from a handful to 125, the employment of some 30 industrial hygienists to work on location in the plants, the addition of a toxicology laboratory employing some 30 toxicologists, and the establishment of a formal risk-assessment unit within corporate headquarters. Such well-developed resources also appear to characterize other comparable large chemical and petrochemical companies, such as Dow, du Pont, Rohm and Haas, Union Carbide, and Monsanto. The presence of 15-20 industrial toxicology laboratories of the size and resources of EPA's most advanced laboratories speaks to an overall industry capability, particularly in light of the fact that a complete toxicology profile for one substance costs around $2.5 million, takes six years to complete, and consumes some 30-50 worker years of effort. Rohm and Haas, for example, logged in approximately 600 samples for toxicological evaluation in 1983 (*Chemecology* 1983).

A 1983 survey of risk activities in chemical companies found that the mean firm surveyed had 84 health and environmental specialists and spent almost 4 percent of its annual sales on toxicity testing and environmental pollution control (Peat, Marwick, Mitchell and Co. 1983, 51-74), figures that have almost certainly increased over the past five years. Nor are extensive capabilities restricted to the chemical industry; Northeast Utilities, employing a staff of 15 professionals in probabilistic risk analysis alone, has one of the most advanced risk assessment units in industry.

But these gains are highly uneven, and many corporations lack the needed hazard management capabilities. Thus, MACHINECORP, although a large multi-national corporation and one of the Fortune 500, had a corporate headquarters health-and-safety staff of three, only one of whom was professionally trained. An internal survey of the chemical industry on environmental auditing in 1983 revealed fully 40 percent of the respondents who did not engage in this rather rudimentary (but critical) health and safety practice, suggesting that the auditing failures evident at Bhopal were not an isolated occurrence (*CMA News* 1983). Other industries engaged in agriculture, mining, and construction or industries dominated by service activities have very poorly developed health and safety functions in comparison with chemicals or nuclear power. And even in the chemical industry, cutbacks in the mid-1980s have rendered the hazard manage-ment functions to be among the first to go.

Finally, the increase in new capabilities at the industrial trade association level is an important component of the revolution in industry hazard manage-ment resources. The Electric Power Research Institute (EPRI) conducts an active program in risk research, and the Institute for Nuclear Power Operations (INPO) is a key institution in evaluating the safety programs of individual utilities and in the training of operators. The Chemical Industry Institute of Toxicology (CIIT) with its $10-million annual budget and staff of 106, funded by 31 chemi-cal manufacturers, has resources nearly 20 percent that of the entire national Toxicology Research and Testing Program, and this is over and above all the toxicology programs at individual corporations. The Chemical Manufacturers Association (CMA) with its staff of 165, augmented by approximately 2,000 representatives from member companies who serve on its committees and panels, has emerged in the 1980s as both a vigorous political lobbyist for the chemical industry and as a major force in chemical risk assessment as well. With its Community Awareness and Emergency Response (CAER) program and its $1-million-per-annum CHEMTREC program (a hotline for information on chemicals in transportation accidents), the CMA extends corporate individual resources substantially. Although more pronounced in these industries, similar industry-wide resources in other parts of the private sector are also commonplace. With 15,000 industrial trade associations in existence, the possibility for indus-try-wide activities in risk assessment and management is substantial. Yet the actual effect of these programs and resources on industrial hazard management as a whole has yet to be assessed and thus remains unknown.

Determinants of hazard-management performance. Ultimately, it is essential to define the key variables that contribute to managerial success or failure. This exploratory study can but propose relevant categories and indicate possible hypotheses. The case studies themselves suggest two major classes of determinants: exogenous and endogenous.

The *exogenous variables* are perhaps the more apparent. It is clear, for ex-ample, that the regulatory systems in health and safety which have been enacted since 1970 have extensively driven the revolution in industrial resources. Whereas all firms have greatly strengthened their hazard-management functions to keep pace with federal and state regulation, external regulation is more important to some industries and corporations than to others. The nuclear industry, for example, is now a leader in assessing catastrophic risk, in emergency planning, in providing defense in depth and redundancy, and in implementing security measures largely because of regulations imposed on a reluctant industry. Those corporations with limited resources and internal organization, such as MACHINECORP, tend to have reactive management programs largely determined by external regulations. Others, such as Rocky Flats and PETROCHEM, have more complex and innovative systems. Although regulation has been essential for achieving the current plateau of industrial hazard-management success, its limitations have also become very apparent.

Liability and insurance costs comprise another exogenous variable, increasingly important and effective over time, even prior to the Bhopal accident. Widespread corporate concern hovers over what industry characterizes as the "deep-pocket" syndrome that persuades juries to make awards according to the ability to pay rather than to the responsibility for the harm that occurred. It has also been argued (Huber 1986) that hazard information rather than hazard severity drives tort litigation. Experiences at Johns Manville, Three Mile Island, and Bhopal have driven home the clear financial vulnerability of corporations to failure in hazard management. In the trucking industry, requisite coverage for a small firm has increased from the traditional $150,000-$300,000 cost of protection to $5 million. Insurance rates, which have been greatly underestimated by insurance companies, have increased over recent years in a number of industries by several hundred to 1,000 percent. In some cases, insurance is difficult to find at any cost. The causes are multiple—courts are making large awards and adding on punitive damages, the basis of claims is being broadened, and environmental clean-up and health costs have skyrocketed. Little of this was predictable to earlier rate-setters in corporate insurance. Small wonder that issues of product liability and insurance are stimulating upgraded hazard management in corporations. Liability, in short, has become a powerful force for improving corporate hazard management.

The effects of hazard liability extend not only to corporate insurance but to the reinsurance market as well, where Lloyds Insurance Group, which has provided much of the reinsurance for the chemical industry, has virtually abandoned the field because of the "unbelievably excessive litigation," high damage awards, and disproportionate defense costs (Baram 1985, 35). By 1987, concern was mounting that fear of litigation was becoming a significant impediment to industrial innovation in the United States (Broad 1987).

Public scrutiny may be another important exogenous factor, as the experience at Rocky Flats (chapter 5) and general societal concern over "toxic" chemicals and radioactive materials indicate. Continuing questions over possible harm and public reviews to detect any failures contribute to a social environment that corporations cannot ignore: "When an industry or company is under intense scrutiny, when it is hot, every operation must be squeaky clean" (Pinsdorf 1987, 105). Union Carbide officials learned this lesson, albeit too late, after blundering through the incident at Institute. Be it the continuing demonstrations and blue-ribbon citizen committees at Rocky Flats, the 1,750 news media queries to Johnson and Johnson following the first Tylenol scare (Kniffin 1987, 22), or the "chemophobia" with which the chemical industry believes it must deal (Cox, O'Leary, and Strickland 1985), the climate of public opinion can be a powerful stimulus for forcing reluctant corporations to upgrade their hazard-management programs.

The *endogenous determinants*, although likely no less significant, are less clear. Rocky Flats, PETROCHEM, and PHARMACHEM all speak to profitability as a key condition correlated with strong hazard-management performance, whereas the deteriorating maintenance and operating situation at the unprofitable Union Carbide plant at Bhopal suggests the reverse. New plants and new technologies, *ceteris paribus*, allow for the incorporation of the latest standards and engineering safeguards. Thus, for example, the layout and construction of Shell Moerdijk, a new plant on a new site, could easily take into account new developments in and prevailing views on health and safety (Shell 1987). The chemical industry has stood at or near the top of safety performance of American industries; the long-established industries, by comparison, tend to have weaker records. Caution is in order, however, for the poor record of waste disposal and occupational exposures in the high-technology industries of Silicon Valley indicates how easily other determinants (high-venture capital/short-term profit horizons) can compromise this generalization.

Commitment of high-level management to health and safety appears to be an important, if somewhat elusive, contributor to effective hazard management. Few corporations boast organizational structures that accommodate the effective use of risk-assessment techniques at higher corporate levels (Rowe 1982, 167). Health and safety issues must penetrate the highest levels of corporate decision making if hazards are to receive primary consideration. Nowhere is this clearer than in Volvo's long-standing emphasis on the innovating and marketing of safety. Moreover, the high placement of health and safety in organizational decision making at PETROCHEM was notably absent at Union Carbide. Lagadec (1987) notes the importance of top-level management's ensuring that hazard problems are not covered up at lower levels and insisting on an effective system of upward flow of information. It is worth noting that such safeguards are difficult to implement in tightly coupled or command-and-control management systems (Perrow 1985).

Rae Zimmerman (1985), in her analysis of industry responses to chemical risks, found market diversity, organizational diversification, and (perhaps) product diversity all relevant to different corporate responses. Again, the size and scale of the corporation seemed important to risk-management programs. Finally, the *degree of hazard* associated with the product itself may drive internal hazard management—a fact very apparent at Rocky Flats but also present at PHARMACHEM. Had managers of the Bhopal plant accorded to methyl isocyanate (MIC) the special handling commensurate to the level of risk, the accident would have been less disastrous.

In short, the case studies offer some evidence to support all seven of these exogenous and endogenous variables. More research is needed, however, to delineate which assume greater importance in what particular situations and contexts.

The shadow government of corporate health and safety. Although not widely recognized, large corporations have an internal regulatory system that "shadows" that of the public sector. In the industrial leaders in health and safety, these are elaborate systems with extensive procedures for identifying and monitoring risks, formal procedures of standard-setting, numerous committees and role specification, mechanisms for conflict resolution, and elaborate means for seeking compliance. This "shadow government" emerges in greatest detail at PETROCHEM where the corporate regulatory system employs three tiers of standards: external (federal regulatory agencies such as OSHA, EPA, and CPSC) and industrial consensus standards, internal (carrying legal implications), and "targets" (more informal and specific objectives). PETROCHEM also allocates substantial emphasis in its regulatory system to maximizing information and minimizing surprises and error. The company shows the same conflict between benefit maximization (or minimization of regulatory burden) and risk minimization that occurs in the public sector, with similar means, both formal and informal, for reaching "balanced" decisions. Moreover, personal "clout" in such decisions appears to have parallels with influence in decision making in the public sector.

Internal industrial regulatory systems are not always well developed, of course, and may be prone to failure. MACHINECORP, despite its size, has a less articulated and formal system and operates with a single tier of externally defined standards. Its audit system is limited and the corporation allocates relatively little effort to assuring compliance. Similarly, the internal regulatory system at the Bhopal plant, which did have a formal audit program, suffered widespread breakdowns in assuring compliance. A strength of the Volvo Corporation, by comparison, is the degree of integration between a well-developed accident feedback program (the accident investigation center) and automobile design and manufacturing.

Anticipatory behavior. Another requisite of effective corporate hazard management is that it be *anticipatory*, in regard both to hazard occurrence and to external regulation, liability, and insurance costs. Historically, the predominant norm for successful industrial hazard management was quick reaction—when problems occurred, the corporation should be able to act swiftly to identify and rectify the situation. The traditional hazard management system has not generally (drugs, agricultural chemicals, and nuclear power are obvious exceptions) been geared to anticipating hazard problems.

The norms are changing. Speedy reaction is no longer the standard of a well-developed corporate hazard management program. Except for Volvo Car Company whose safety standards have been well ahead of national public regulations, corporations generally are only now beginning to engage in anticipatory hazard control. Rocky Flats, which is distinctly anticipatory in its behavior, has lowered its radiation exposure standard to 20 percent of that allowed by Nuclear Regulatory Commission (NRC) regulations. PETROCHEM has adopted a number of internal standards more stringent than those required by OSHA, seeks to anticipate external regulation by altering its productive processes and lowering (in cost-effective manner) occupational exposures, and by the mid-1980s was conducting prospective epidemiological studies to identify potential hazards.

Monsanto has been a leader in actions to enhance community knowledge of chemical hazards and to increase local emergency-response programs. The 3-M Corporation has pioneered innovative programs in waste reduction, achieving 50 percent waste reductions between 1975 and 1986 (Koenigsberger 1986). Independent of regulatory requirements, liability and insurance are driving stronger anticipatory hazard management, a situation that poses new challenges to industrial organization but which may also reduce the role of governmental regulation.

Low-probability risks and defense in depth. Low-probability, catastrophic risks pose inherent difficulties for corporate hazard management. Where experience provides little guidance and actuarial indicators of risk are unavailable, formal probabilistic risk analysis must substitute for experience. Such assessment requires analytic resources not available in most corporations. Recently, when the authors visited a modern computer chip plant, programs for handling routine risks appeared well developed, formally stated, and enforced. Questioning revealed, however, that low-probability events were not in the consciousness of corporate health and safety managers and had not undergone formal (or even informal) assessment.

Defense in depth—multiple layers of safety systems incorporating redundancy—is well developed in certain industrial sectors in which issues of catastrophic risk have been the focus of scientific and public controversy. Rocky Flats has *eleven* levels of defense in depth; nuclear plants have similar well-developed layering and redundancy. But Bhopal, Three Mile Island, Rocky Flats, the space-shuttle Challenger, and Chernobyl all speak persuasively to the vulnerability of such systems to compromise by shoddy maintenance, faulty design, management failure, and human error. The case studies deliver a somewhat ambiguous message—defense in depth is a sound safety philosophy but its success requires formal assessment (beyond much current corporate capability) for its emplacement, a willingness to invest heavily in safety, and vigilant monitoring and implementation. Paul Shrivastava (1987, 132) cautions that defense in depth includes redundancy in personnel as well as in engineered safety systems.

An Agenda of Needed Research

The results of our studies point in three major directions for future research on the corporate management of hazards. The first is a need to broaden current understandings by moving from individual cases to comparative studies. This could treat the overall structure and determinants of industrial hazard management, including analyses of entire industrial sectors as well as individual firms, and extending the comparative perspective to other industrialized countries and the developing world. The second direction recognizes that exploratory ventures, such as those undertaken for this book, cannot achieve the depth requisite for a full understanding of the inner workings of corporate hazard management. Accordingly, it is essential to initiate projects that probe the internal dynamics of management systems and decision-making processes. The third direction is to follow up in detailed study some of the opportunities for improving hazard management by integrating better the private resources of corporations into overall efforts of the public sector.

The determinants of hazard management. The case studies in this book suggest wide variation, or substantial unevenness, in how corporations manage hazards. In general, the authors had access to large, wealthy corporations that took pride in their programs for managing hazardous products. Building upon this experience in learning how to characterize the hazard-management commitment, function, organization, and resources of corporations, further research should undertake a more systematic exploration of how these characteristics vary by hazard and by corporation size, age, profitability, planning horizon, market and product diversity, and organizational structure. Does depth in hazard management increase with hazardous products and processes, large size and plentiful resources, profitability, and experience? What role does tight and loose coupling play in the response of corporate management systems to hazard surprises? Is it true, as is commonly suggested, that much of the failure in industrial hazard management is dominated by small, marginal firms that cannot, will not, or do not, invest in safety and health protection? Since the studies reported on in this book did not include such firms (although we tried—access to such firms is very difficult), the results provide greater insight into success than into sources of failure. Systematic study is needed to analyze the behavior of a spectrum of firms within a given industry.

Management convergence. Modern hazard management employs a wide variety of technological and behavioral controls designed to prevent or reduce hazard. In contrast with the stereotypes, it is apparent that safety is not always in conflict with other corporate goals. The experience with nuclear power, and at Volvo, indicates that a well-developed quality assurance program enhances *both* reliability and safety. Effective waste-reduction programs can result in actual cost savings and materials recovery while protecting the corporation from uncertain environmental or health liability claims. Thus opportunities exist for structuring hazard management to converge with production or product management goals and to afford long-term income protection for the corporation. Such hazard controls, as our PHARMACHEM example clearly indicates, reduce the threat of accidents that interrupt production or that contaminate products even as they protect the health and safety of workers and consumers. Other controls, however, do not serve any inherent dual purpose and are employed only under the demands of external regulation, or the threat of a lawsuit or an insurance bill. Can the convergences and divergences between profit and health and safety be more clearly identified? Can ways be found to encourage the convergences and reduce the conflicts?

Trade unions and hazard management. Traditionally, trade unions have not been a strong force for protecting the health and safety of workers in the United States. Despite participation in corporate health and safety committees and advocacy in particular hazard cases, the union presence has been anemic overall and other priorities have generally dominated union actions. Over the past decade, however, some unions, such as the Oil, Chemical and Atomic Workers International Union and the United Steelworkers of America, have become decidedly more aggressive on health and safety questions. Some union contracts, for example, are now stipulating corporate health and safety commitments and programs. But, as a 1984 survey has made clear, U.S. trade unions have very limited resources and only modest programs (Wolfe and Abrams 1984). Except at Volvo in Sweden (chapter 4), the preceding case studies found no substantial union role in hazard management. Meanwhile, the trade union movement in the United States has faltered during the 1980s. A more searching assessment needs to address the changing role of unions and the means of increasing their effectiveness in occupational hazard management.

National comparisons. Comparative studies of hazard management in industrialized countries have suggested a greater reliance on corporations and trade unions in hazard management and more confidence in regulatory discretion outside the United States. Some evidence indicates that other countries are heavily reliant upon U.S. risk assessments but quite efficient, with fewer resources, in setting standards and informally negotiating compliance in a flexible manner with industry. Whereas some studies (Brickman, Jasanoff and Ilgen 1985; Kelman 1981, Pijawka 1983) conclude that the results are quite similar despite the varied approaches, other studies (e.g., Kasperson 1983) suggest that the outcomes may be significantly different.

The case studies presented in this book are only suggestive. The selection of Volvo as a case study allowed for specific contrast with the U.S. automobile industry in the development and marketing of safety, and the comparison certainly suggests that the American automobile industry has things to learn from Volvo. A systematic set of national comparisons of industrial hazard management is needed to address such questions as: Do Europeans and Japanese trust their corporations more than Americans do? If so, is this trust justified? Do they make better public use of corporate resources? Does the consensual, negotiated approach to standard-setting and implementation produce higher or lower levels of safety and compliance? What is the role of the trade unions in industrial hazard management? What novel mechanisms have been developed as alternatives or complements to regulation? How transferable are hazard management success across political cultures and economies? In short, how well does successful hazard management travel?

Corporate social-responsibility programs. In the wake of the environmental and civil rights movements, U.S. corporations have widely adopted social responsibility programs. Such programs generally provide the philosophical underpinnings for and include stipulations affecting corporate hazard management. Thus, the product-stewardship program at Dow Chemical has for years recognized an obligation for "cradle-to-grave" care of chemicals (Buchholz, Evans, and Wagley 1985). A social-responsibility program explicitly commits PETROCHEM to a goal of leadership in health and safety protection in the petrochemical industry. A similar program was credited by Johnson and Johnson with playing an important role in the Tylenol scare (Kniffin 1987). but limited knowledge exists as to how such programs really function, the extent to which they make a difference in the corporation, the methods for verifying compliance, and their long-term impacts on managerial behavior. At the reconnaissance visit to Dow in the late 1970s, company responses to the authors' queries indicated that despite a well-articulated program and extensive investments, independent validation of the implementation and compliance achieved had not occurred. Post-mortems on the Bhopal catastrophe attest to similar defects in verifying implementation. A careful study of whether these corporate ethical principles have actually affected corporate objectives and the behavior of company managers or whether they are simply window-dressing would substantially add to current knowledge.

Auditing the audits. Formal auditing programs have been widely adopted to assure compliance with corporate objectives in hazard management (as in the stated goal for lowering the occupational injury rate). Such audits are mounted internally in most large corporations; smaller firms often turn to consulting firms (e.g., Arthur D. Little) that specialize in such activity or depend upon the review of insurance companies. The ambition, scope, and rigor of audits vary enormously as do the resources allocated in their behalf, the frequency of application, the degree of integration with incentives to assure compliance by plant managers, and the corporation's determination to verify compliance. At Bhopal, it was apparent that the auditing procedure lacked teeth and follow-through. The Chemical Manufacturers Association has established auditing guidelines to improve on the performance of member firms, 40 percent of which in 1983, neither had such programs in place nor claimed any intention to develop them (*CMA News* 1983, 12). A rigorous analysis of the scope and depth of such audits, their respective content, measures of success and failure, and effectiveness in correcting weaknesses and problems would be extremely valuable.

Hazard communication and worker participation. Worker exposure to risk, it is sometimes argued, is voluntary or semi-voluntary, based upon workers' knowledge of the hazards to which they are exposed and the degree to which they are compensated to take risks. Whereas it is apparent that the growth in industry programs designed to communicate risks to workers (Melville 1981; Viscusi 1984, 1987) and the OSHA rule for hazard communication (OSHA 1983) are major developments, evidence suggests that worker understanding of risk is often very inadequate (Nelkin and Brown 1984). Whereas Rocky Flats speaks to a relatively high level of understanding of radiation hazards, Indian workers at Bhopal had a totally inadequate level of knowledge. Meanwhile, Sweden provides a telling case of much higher levels of ambition and program development (also not, however, rigorously evaluated as to impact). Although the communication of risk is currently fashionable in the Washington regulatory community as well as among scholars, relatively little is known about risk communication in the workplace. A searching examination of worker knowledge of hazards, the effectiveness of the hazard communication rule as it is implemented, the attributes of successful programs, and how such knowledge can be utilized to improve worker participation in identifying and reducing risks would greatly enhance industrial hazard-management efforts.

The culture of safety. Corporate officials involved in hazard management speak persuasively of the "culture of safety" that accompanies upgraded safety. Recently, a representative of an oil company described how it happened at his corporation: as a result of an industrial accident, the Chief Executive Officer proclaimed a corporate commitment to substantially increased levels of safety. The enactment of new programs required the breakdown of traditional ways of thinking, the encouragement of new attitudes. The effort consumed much of the corporation's management energy over a three-year period but did make major progress, in this person's view, for creating this new culture.

The corporation most frequently cited for having such a culture is du Pont, where aggressive hazard management programs have spanned a 15-year period. A different picture emerges in the Bhopal case, where understanding of toxic materials and the need for preventive maintenance was sadly lacking. How does a "culture of safety" arise and how it is maintained? What is its content? Does it transcend changes in executive officers or crises in profitability or product line? Is it transferable to new locations, does it survive divestment, and is it teachable in business schools?

For much of the past decade, the low-level, chronic health hazards have dominated thinking and attention in industrial hazard management. The past disasters at Three Mile Island (an economic disaster), Seveso (an ecological disaster), Bhopal (a human disaster), Chernobyl (a human and ecological disaster), Mexico City (a human disaster), and the Rhine pesticide contamination (an ecological disaster) have fundamentally changed the corporate hazard agenda. The case studies in this volume treat two cases of catastrophic risk management— Rocky Flats, where the catastrophe has thus far been averted, and Bhopal, where it occurred. The management of catastrophic risk is very demanding, requiring extensive analytic resources, meticulous corporate attention, and well-developed data bases. It is undoubtedly most advanced in the defense sector and in nuclear power plants. The use of probabilistic risk analysis has increased dramatically in the nuclear industry over the past decade (Kasperson and Kasperson 1987)—some 20 nuclear power plants now have or are in the process of preparing site-specific studies (Kasperson et al. 1987). Bhopal made it clear that the chemical industry faces low-probability/catastrophic risks as well and must incorporate relevant policies for assessing and managing low-probability risks. Other industries, such as biotechnology, may yet experience these problems. What is the state of catastrophic-risk assessment in industry? To what extent are generic means for preventing catastrophic risk—"defense in depth," equipment redundancy, advanced accident analysis, emergency training, remote siting, the use of simulators— routinely employed in other industries? How serious are the gaps?

Public use of private resources. Ideally, the hazard makers are potentially the best hazard managers. But inherent conflicts of interest and frequent failures undermine the potential. Are there societal mechanisms for reducing these conflicts of interests? Are these mechanisms adaptable to corporate hazard management? Are there analogs in decentralized review mechanisms (e.g., biohazard review committees with public members, professional certification in engineering)?

A National Research Council (1984) report, based on a sample of 53,500 distinct chemical entities in commercial products, found that minimal toxicity data in publicly accessible sources existed for only a third of drugs and pesticides, a quarter of cosmetics, and a fifth of chemicals in commerce. Yet large amounts of high-quality toxicity data not reportable under TSCA and not available for public use reside in corporate files. Corporate officers characteristically raise numerous objections and obstacles to sharing such data: the data may not be useful or comparable; they reveal research and development strategies of the firm; toxicity testing is a major cost of product development and corporations compete in this activity; firms might be liable for the information provided, and so forth. Can ways be found to meet the reasonable concerns while affording greater use of this major resource?

Hybrid institutions. A growing number of novel hazard-management institutions combine industry, public interest groups, and government. The Chemical Manufacturers Association has joined in negotiated rulemaking (as with inadvertent releases of PCBs), in joint suits with environmentalists (as in hazardous waste health effects), in cosponsoring independent studies (of hazardous waste), and in broad-based efforts for nonregulatory solutions to hazardous waste clean-up (Clean Sites, Inc.). The 1980s have also witnessed efforts to find new ways to involve the public in hazard control—the Ruckelshaus efforts to involve the public in risk assessment at the Tacoma shelter and the current EPA efforts to communicate the risks of indoor radon to homeowners are examples. Environmental mediation has been previously used with effectiveness by the Keystone Center in dealing with radioactive wastes.

Another class of hybrid is mandated by regulation such as the so-called "squeal law" aspects of TSCA that mandate the inclusion of private information into the public sector. A systematic effort to conceptualize such opportunities, to evaluate outstanding examples, and to assess their future promise might answer such questions as: What is the potential of hybrid public/private hazard management to provide cost-effective hazard reduction? Is voluntary corporate action a viable alternative to regulatory mandate? What are the dangers of collaborative industry/environmental initiatives that substitute for the normal regulatory process? Are there situations where mediation is preferable to regulation? Where it is not viable?

The Future of Corporate Hazard Management

Hazard management in corporations simultaneously functions on three time horizons: legacies, usually grim, from the past; ongoing management of present production processes and products; and the development of future processes and products. Much of the corporate hazard management effort directed to the past is defensive or compensatory, trying to correct, cover-up, or forget hazards created in the past. The present activity is both complex and variable among corporations. Corporations currently are defending products, reducing hazards where feasible, complying with regulations while resisting future ones, seeking to prevent disasters, identifying hidden hazards, and avoiding liability wherever possible. At the same time, some corporations are seeking safer products and improved engineered safety in new or rebuilt plants whereas others cling to past practices that endanger workers and publics.

Because all three levels of activity are ongoing and involve very different kinds of responses and resources, the public image and private practice of corporate hazard management are confounded. The cross-fires over past decisions and newly proposed regulations should not, in our view, consume all of society's attention to corporate hazard management. The most pressing existing need is to decrease the great disparities in corporate hazard management, to bring all corporations in a particular industrial sector to the standard of the best—to spread, in short, the best of current corporate practice throughout the private sector. A focus on the future is also needed—to examine how new products and new plant designs can better reduce hazards, how cultures of safety germinate and are nourished, how industry can be made more socially responsible, and how industrial capabilities can be more effectively used in behalf of the larger societal good. These in turn call for embedding the lessons of the past—in new technology development, in corporate organizational structures, in venture capital, and in the new constantly arising hazards whose impacts are imperfectly understood but whose benefits continue to be needed.

References

ACGIH (American Conference of Governmental Industrial Hygienists). 1984. *TLVs: Threshold limit values for chemical substances and physical agents in the work environment and biological exposure indices with intended changes for 1984-1985.* Cincinnati: ACGIH.

Acquavella, J.F., *et al.* 1981. *Lost-to-followup bias in occupational analysis.* Los Alamos: Los Alamos Scientific Laboratory.

AIChE (American Institute of Chemical Engineers). 1987. *Guidelines for hazard evaluation procedures* New York: AIChE.

Arkin, W.H., T.B. Cochran, and M. Hoenig. 1982. The U.S. nuclear stockpile. *Arms Control Today* 12(4):1.

Baram, Michael. 1985. Chemical industry hazards: Liability, insurance and the role of risk analysis. Unpublished paper presented to the Joint Conference on the Transportation Storage, and Disposal of Hazardous Materials. Laxenburg, Austria. International Institute for Applied Systems Analysis, July 1-6. [Note: For a slightly different version of this paper, see Baram (1987), below.]

Baram, Michael. 1987. Chemical industry hazards: Liability, insurance and the role of risk analysis. In *Insuring and managing hazardous risks: From Seveso to Bhopal*, ed. Paul R. Kleindorfer and Howard C. Kunreuther, 415-441. New York: Springer-Verlag.

Bean, E.W. 1976. Attachment to July 2, 1976 letter to J.F. Burke, *Cost of May 11; 1969 Fire at Rocky Flats*, as quoted by S. Chinn 1981.

Bohlin, N. 1967. A statistical analysis of 28,000 accident cases with emphasis on occupant restraint value. *SAE 67, 925.* Warrendale, PA: Society of Automotive Engineers.

Bowonder, B., Jeanne X. Kasperson, and Roger E. Kasperson. 1985. Avoiding future Bhopals. *Environment* 27 no. 7 (September):6-13, 31-37.

BRCC (Blue Ribbon Citizen's Committee) 1983. *Final report on the Long-range Rocky Flats utilization study.* Denver: Colorado Governor's Office.

Brickman, Ronald, Sheila Jasanoff, and Thomas Ilgen. 1985. *Controlling chemicals: The politics of regulation in Europe and the United States.* Ithaca, NY: Cornell University Press.

Broad, William C. 1987. Does the fear of litigation dampen the drive to innovate? *New York Times* (12 May):C1 and C9.

Brushnan, Bharat, and Arun Subramanian. 1985. Bhopal: What really happened? Special Report 1, *Business India* 7, Issue 82 (March).

Buchholz, Rogene A., William D. Evans, and Robert A. Wagley. 1985. Dow Chemical and product stewardship. Chapter 6 in *Management response to public issues: Concepts and cases in strategy formation*, 99-111. Englewood Cliffs, NJ: Prentice Hall.

Carson, Rachel. 1962. *Silent spring.* Boston: Houghton Mufflin.

Cathcart, Chris. 1987. Comments delivered to the Committee on Risk Communication and Risk Perception, National Research Council, July 23.

Chemecology. 1983. Toxicology testing helps company assess chemical, product safety. *Chemecology* (March):7.

Chinn, S. 1981. *The relation of the Rocky Flats plant and other factors to 1969-71 cancer incidence in the Boulder area.* Denver: Fairfield and Woods.

CMA (Chemical Manufacturers Association). 1987. *CAER progress report 1986.* Washington: CMA.

CMA News. 1983. CMA unveils position paper on environmental auditing: Majority of member companies use audits or intend to. *CMA News* 11 no. 8:12.

Cox, Geraldine V., Timothy F. O'Leary, and Gordon D. Strickland. 1985. The chemical industry's view of risk assessment. In *Risk Analysis in the private sector*, ed. Chris Whipple and Vincent T. Covello, 271-284. New York: Plenum.

Delhi Science Forum. 1985. Bhopal gas tragedy. *Social Scientist* (Kerala) 13 (No. 140).

Denver Post. 1983. Antinukery vs. science (editorial). 5 February.

Diamond, Stuart. 1985. The disaster in Bhopal: Lessons for the future. *New York Times* (3 February):8.

DOE (U.S. Department of Energy). 1980. *Rocky Flats plant site: Final environmental impact statement.* DOE-EIS-0064. Washington: DOE. 3 vols.

DOE (U.S. Department of Energy). 1983. *Long-range Rocky Flats utilization study.* ALO-1983. Albuquerque: DOE.

DOE (U.S. Department of Energy). 1988. *Technical safety appraisal of Building 776/777, Rocky Flats Plant.* DOE EH-0061. Washington: Environment, Safety, and Health, Office of Nuclear Safety, DOE, March.

Financial Times (London). 1984. Union Carbide halts production of pesticide use. December 5, p. 1.

GAO (General Accounting Office). 1987. *Nuclear materials. Alternatives for relocating Rocky Flats Plant's plutonium operations.* GAO RCED-87-93. Washington: GAO.

Haddon, W., Jr. 1972. A logical framework for categorizing highway safety phenomena and activity. *Journal of Trauma* 12:193-207.

Hohenemser, Christoph. 1987. Public distrust and hazard management success at the Rocky Flats nuclear weapons plant. *Risk Analysis* 7 (June):243-259.

Hohenemser, Christoph, Robert W. Kates, and Paul Slovic. 1983. The nature of technological hazard. *Science* 220:378-384.

Hohenemser, Christoph, and Ortwin Renn. 1988. Shifting public perceptions of nuclear risk: Chernobyl's other legacy. *Environment* 30 no. 3 (April):4-11, 40-45.

Holusha, John. 1987. U.S. fines Chrysler $1.5 million, citing workers' exposure to peril. *New York Times,* 7 July, pp. A1, A17.

Huber, Peter. 1986. The Bhopalization of American tort law. In *Hazards: Technology and fairness,* ed. National Academy of Engineering, 89-110. Washington: National Academy Press.

Ingelstam, L. 1980. Decision-making and strategic social planning. *Acta Psychologica* 45:265-271.

Jaksch, F.O. 1979a. Driver-vehicle interaction with respect to steering control ability. *SAE Technical Paper Series* 790740. Warrendale, PA: Society of Automotive Engineers.

Jaksch, F.O. 1979b. Vehicle characteristics describing the steering control quality of cars. In *Seventh International Technical Conference on Experimental Safety Vehicle..., Paris, June 5-8, 1979,* 815-846. DOT HS-805 199. Washington: Department of Transportation, National Highway Traffic Safety Administration.

Jaksch, F.O., R. Gustafsson, and L. Solberg Larsen. 1974. Volvo's safety system integration in producing automobiles: Crash avoidance engineering. *FISITA XIV Congress International,* Paris (12-17 May).

Johnson, Branden B. 1985. Tales of woe: A literature survey. In *Perilous progress: Managing the hazards of technology,* ed. Robert W. Kates, Christoph Hohenemser, and Jeanne X. Kasperson, 291-310. Boulder, CO: Westview Press.

Johnson, C.J. 1981. Cancer incidence in an area contaminated with radionuclides near a nuclear installation. *Ambio* 10(4)176-182.

Kail, L.K., J.M. Poulson, and C.S. Tyson. 1982. *Operational safety survey.* Danbury, CT: Union Carbide Corporation.

Kasperson, Roger E. 1983. Worker participation in protection: The Swedish alternative. *Environment* 25 (No. 4):13-20, 40-43.

Kasperson, Roger E., James E. Dooley, Bengt Hanson, Jeanne X. Kasperson, Timothy O'Riordan, and Herbert Paschen. 1987. Large-scale nuclear risk analysis: Its impacts and future. In *Nuclear risk analysis in comparative perspective: The impacts of large-scale risk assessment in five countries,* ed. Roger E. Kasperson and Jeanne X. Kasperson, 219-236. Risks and Hazards Series, 4. Boston: Allen and Unwin.

Kasperson, Roger E., and Jeanne X. Kasperson, eds. 1987. *Nuclear risk analysis in comparative perspective: The impacts of large-scale risk assessment in five countries.* Risks and Hazards Series, 4. Boston: Allen and Unwin.

Kates, Robert W. 1978. *Risk assessment of environmental hazard.* SCOPE Report 8. Chichester: John Wiley & Sons.

Kates, Robert W., Christoph Hohenemser, and Jeanne X. Kasperson, eds. 1985. *Perilous progress: Managing the hazards of technology.* Boulder, CO: Westview Press.

Kelman, Stephen. 1981. *Regulating America, regulating Sweden: A comparative study of occupational safety and health policy.* Cambridge, MA: MIT Press.

Keswani, Rajkumar. 1982a. Sage, please save this city. *Saptahik Report* (17 September).

Keswani, Rajkumar. 1982b. Bhopal on the mouth of a volcano. *Saptahik Report* (1 October).

Keswani, Rajkumar. 1982c. If you don't understand you will be wiped out. *Saptahik Report* (8 October).

Keswani, Rajkumar. 1984. Bhopal's killer plant. *Indian Express* (Delhi) (9 December).

Kniffin, F. Robert. 1987. Corporate crisis management. *Industrial Crisis Quarterly* 1 (Spring):19-23.

Koenigsberger, M.D. 1986. Paper presented at Governor's Conference on Pollution Prevention Pays. Nashville, TN. March.

Krey, P.W. 1974. Plutonium 239 contamination in the Denver area. *Health Physics*. 46:117.

Krey, P.W. 1976. *Plutonium and Americium contamination in Rocky Flats soil-1973*. HASL-304. Washington: Health and Safety Laboratory, Energy Research and Development Administration.

Krey, P.W., and E.P. Hardy. 1971. *Plutonium in soil around the Rocky Flats Plant*. HASL-235. Washington: Health and Safety Laboratory, U.S. Atomic Energy Commission.

Lagadec, Patrick. 1982. *Major technological risk: An assessment of industrial disasters*. New York: Pergamon Press.

Lagadec, Patrick. 1987. From Seveso to Mexico and Bhopal: Learning to cope with crises. In *Insuring and managing hazardous risks: From Seveso to Bhopal and beyond*, ed. P. R. Kleindorfer and H. C. Kunreuther, 13-46. New York: Springer-Verlag.

Larsen, L.S. 1975. Systems technology in support of road safety legislation. *Road Safety Symposium*, Cape Town, September 8-10.

Lawrence, R.M., et al. 1984. *Report of the Rocky Flats employees health assessment group*. Denver: Governor's Science and Technology Advisory Council.

Lepkowski, Will. 1985. Special Report: Bhopal. *Chemical and Engineering News* 63 (No. 48):18-32.

Melville, Mary. 1981. Risks on the job: The worker's right to know. *Environment* 23 (No. 9):12-20, 22-45.

National Research Council. 1980. Committee on the Biological Effects of Ionizing Radiations (BEIR). *The effects on populations of exposure to low levels of ionizing radiation: 1980.* Washington: National Academy Press.

National Research Council. 1984. Steering Committee on Identification of Toxic and Potentially Toxic Chemicals for Consideration by the National Toxicology Program. *Toxicity testing: Strategies to determine needs and priorities.* Washington: National Academy Press.

Nelkin, Dorothy, and Michael Brown. 1984. *Workers at risk: Voices from the workplace.* Chicago: University of Chicago Press.

NRC (U.S. Nuclear Regulatory Commission). 1975. *Reactor safety study.* WASH-1400, NUREG 75/014. Washington: NRC.

NRC (U.S. Nuclear Regulatory Commission). 1984. *Probabilistic risk assessment (PRA) reference document: Final report.* NUREG-1050. Washington: NRC.

Nuclear Waste News. 1987. Rocky Flats trial burn poses little health risk, CDC report contends. *Nuclear Waste News* 7 (13 August):232.

OSHA (Occupational Safety and Health Administration). 1983. Hazard communication: Final rule. *Federal Register* 48 no. 228 (25 November):53280-53348.

OTA (Office of Technology Assessment). 1983. *Technologies and management strategies for hazardous waste control.* OTA-M-196 (March). Washington: Government Printing Office.

Peat, Marwick, Mitchell and Co. 1983. *An industry survey of chemical company activities to reduce unreasonable risk.* Final report, prepared for the Chemical Manufacturers Association. February 11. Washington: The Author.

Pijawka, K. David. 1983. A comparative study of the regulation of pesticide hazards in Canada and the United States. Ph.D. dissertation, Clark University, Worcester, MA.

Perrow, Charles. 1985 *Normal accidents: Living with high-risk technologies.* New York: Basic Books.

Pinsdorf, Marion. 1987. *Communicating when your company is under siege: Surviving public crisis.* Lexington, MA: Lexington Books.

Poet, S.E., and E.A. Martell. 1972. Plutonium-239 and Americium-241 contamination in the Denver area. *Health Physics* 23:537-548.

Putzier, E.A. 1981. *Impact on Rocky Flats Plant of proposed reduction in occupational radiation exposure standards.* Golden, Colorado: Rockwell International.

Qureshy, Ahtesham. 1985. Did Buch okay Carbide plant site? *Hindustan Times* (13 February):1,8.

Ramaseshan, Radhika. 1984. Government responsibility for Bhopal gas tragedy. *Economic and Political Weekly* 19 (15 December):2109.

Ramaseshan, Radhika. 1985. Bhopal gas tragedy: Callousness abounding. *Economic and Political Weekly* 20 (12 January):56-57.

Reyes, M., *et al.* 1984. Brain tumors at a nuclear facility. *Journal of Occupational Medicine.* 26(10):721-724.

Rijnmond Public Authority. 1982. *Risk analysis of six potentially hazardous objects in the Rijnmond area, a pilot study: A report to the Rijnmond Public Authority.* Boston, MA: A. Reidel.

Rockwell International. 1981. Press Release, October 1981. Denver, CO: Rockwell International.

Rowe, William D. 1982. *Corporate risk assessment.* New York: Marcel Dekker.

Shell Internationale Petroleum Maatschappif, B.V. 1987. Shell Moerdujk: A modern facility. *Industry and Environment* 10 No. 1 (January/February/March):11-14.

Shrivastava, Paul. 1987. *Bhopal: Anatomy of a crisis.* Cambridge, MA: Ballinger.

Sorensson, P.A. 1978. Internal Volvo communication.

Sorensson, P.A. n.d. *Riktlinjer for kvalitetsstyrning: Leverantors information fran AB Volvo.* (Frame of reference for quality control: Information to subcontractors from AB Volvo). This information is also available in English. Göteborg: Volvo.

Starr, Chauncey. 1969. Social benefit versus technological risk. *Science* 165:1232-1238.

Starr, T.B. 1987. A new initiative: The CIIT program on risk assessment. *CIIT Activities* 7 (No. 3):1,5.

Swedish Department of Communication. 1983. *Trafiksakerhet-problem och atgarder: Betankande avgivet trafiksakerhetsutredningen.* Stockholm: The Department.

U.K. Health and Safety Executive. 1978. *Canvey: An investigation of potential hazards from operations in the Canvey Island/Thurrock area.* London: HMSO.

U.K. Health and Safety Executive. 1981. *Canvey: A second report: A review of potential hazards from operations in the Canvey Island/Thurrock area three years after publication of the Canvey report.* London: HMSO.

UNEP (United Nations Environment Programme). 1988. Hazardous chemicals. *UNEP Environment Brief* no. 4. 8 pp. (unnumbered).

Union Carbide Corporation. 1985. *Annual report 1984.* Danbury, CT: Union Carbide Corporation.

Union Carbide Corporation Engineering and Technology Services. 1984. *Operational safety/health survey: Institute MIC II unit.* Charleston, W. VA: Union Carbide Corporation.

Viscusi, W. Kip. 1984. *Risk by choice: Regulating health and safety in the workplace.* Cambridge, MA: Harvard University Press.

Viscusi, W. Kip. 1987. *Learning about risk: Consumer and worker responses to hazard information.* Cambridge, MA: Harvard University Press.

Voelz, G.L., *et al.* 1981. *An update of epidemiologic studies of plutonium workers.* LA-UR-82-123 (revised). Los Alamos, NM: Los Alamos Scientific Laboratory.

Volvo. 1967. *Report on a statistical analysis of 28,000 accident cases.* Göteborg: Volvo.

Washington Post. 1988. Bhopal death toll put at 2,998. 29 April, p. A26.

Weinberg, Alvin. 1972. Science and trans-science. *Minerva* 10:209-222.

Wilkenson, G.S., *et al..* 1983. *Mortality among plutonium and other workers at a nuclear facility.* Proc. of the midyear Topical Meeting of the Health Physics Society, CONF-830101. January 9-13, pp. 328-337.

Index

ACGIH. *See* American Conference of Governmental Industrial Hygienists (ACGIH)
American Conference of Governmental Industrial Hygienists (ACGIH), 51
American Petroleum Institute, 19
Anticipatory behavior, 20, 55, 64-65
Audits, 16, 21-25, 51, 105, 107, 122, 128
Automobile safety, 7, **57-78**

BEIR report, 91
Bhopal, 98, **101-117**, 120, 123, 124, 129
Blue Ribbon Citizens Committee, 82, 93
Bohlin, N., 66

CAER program, 116, 122
Canvey Island, 117
Carson, Rachel, 1
Catastrophic hazards, 93-97
CENTED, *See* Center for Technology, Environment, and Development (CENTED), Clark University
Center for Chemical Process Safety, 117
Center for Technology, Environment, and Development (CENTED), Clark University, 3, 9
Centers for Disease Control, 82
Chemical industry, 3,101-102, 115, 122
Chemical Industry Institute of Toxicology (CIIT), 117, 122

Chemical Manufacturers Association, 4, 5, 7-9, 116, 122, 129, 131
CHEMTREC, 122
Chernobyl, 97, 103, 113, 126, 130
CIIT. *See* Chemical Industry Institute of Toxicology (CIIT)
Clark University, 3, 9
Community Awareness and Emergency Response (CAER) program. *See* CAER program
Consumer Product Safety Act, 30, 31
Corporate goals, 2, 7, 17, 18, 45-46, 52-53, 55, 105, 114, 117, 120, 128
Corporate resources, 6, 8, 16, 17, 19, 50, 53, 58-63, 69, 104-105, 117, 121-123
Cubatão, São Paulo, Brazil, 103, 106
Culture of safety, 129-130

Defense in depth, 7, 97-98, 113, 114-115, 126
Department of Energy (DOE), 79, 83, 85, 93
Digital Equipment Corporation, 4, 5
Disasters (list), 103
Dow Chemical Company, 121, 128
du Pont Chemical Company, 121, 130

Electric Power Research Institute (EPRI), 122
Emergency preparedness, 115
EPRI. *See* Electric Power Research Institute

141

Exclusion zones, 79, 113, 114

Federal Highway Administration
    (U.S.), 69
Federal Insecticide, Fungicide, and
    Rodenticide Act (FIFRA), 30
Feedback system, 70-74, 76
FIFRA. *See* Federal Insecticide,
    Fungicide, and Rodenticide Act
    (FIFRA)

Gyllenhammar, Pehr G., 63

Hanford, Washington, 79, 84
Hanover Insurance, 5
Hazard Assessment Group, Clark
    University, 3
Hazard communication, 50, 107,
    115, 123, 129
Hazard profiles, 105
Health-surveillance systems, 33,
    34, 55
Healthy-worker syndorme, 91, 92-
    93
Highway Safety Act, 64

Indian Council of Agricultural Re-
    search, 106
Industrial disasters (list), 103
Information, 50, 107, 115, 123,
    129
INPO. *See* Institute for Nuclear
    Power Operations (INPO)
Institute, West Virginia, 107, 115,
    123
Institute for Nuclear Power Opera-
    tions (INPO), 122
Insurance, 75, 123

Jaksch, F.O., 68
Johnson and Johnson, 123, 128

Keswani, Rajkumar, 107

Larsen, Lauritz Solberg, 64
Lawrence, Robert, 93
Liability, 32, 33, 69, 74-75, 123

Lloyds Insurance Group, 173
*Long-range Rocky Flats utilization
    study*, 82, 93, 94, 98

MACHINECORP, 4, 5, 6, 122, 123,
    125
Material safety data sheets (MSDSs),
    32, 51
Matrix management, 25-26, 27
Methyl isocyanate (MIC), 102, 106,
    109, 111-112, 115
Mexico, 103, 106
MIC. *See* Methyl isocyanate (MIC)
Monsanto Chemical Company, 4,
    5, 121
Motor Vehicle Safety Act, 64
MSDSs. *See* Material safety data
    sheets (MSDSs)
Multinational corporations, 117

National Highway Safety Bureau
    (U.S.), 69
National Institute for Occupational
    Safety and Health (NIOSH), 51
National Research Council (U.S.),
    91, 130
Northeast Utilities, 121
Nuclear Regulatory Commission,
    116, 125

Occupational hazards, 32-37, 39,
    43-56, 87-93
Occupational Safety and Health
    Administration (OSHA), 22-23,
    50, 51, 55, 121, 125, 129
OSHA. *See* Occupational Safety
    and Health Administration
    (OSHA)

PELs. *See* Permissible exposure
    limits (PELS)
*Perilous progress: Managing the
    hazards of technology*, 9
Permissible exposure limits (PELS),
    22-23
PETROCHEM Corporation, 4, 5, 6,
    7, 8, **15-41**, 43, 45, 50, 60,

PETROCHEM Corporation (*contd.*),
63, 81, 87, 104, 105, 120,
121, 123, 124, 125
PHARMACHEM Corporation, 4, 5,
6, 7, **43-56**, 63, 87, 120,
124, 127
PRA. *See* Probabilistic risk
assessment (PRA)
Probabilistic risk assessment
(PRA), 94-97, 116-117, 121
Public opinion, 98, 99, 123-124

Quality control, 5, 58, 63, 69-72,
74

Reactor safety study, 94, 96, 97
Recalls, 62, 75
Recordable incidence rate, 36, 38
Remote siting, 79, 113, 114
Risk acceptability, 24, 52, 53
Risk communication, 50, 107,
115, 123, 129
Risk-region approach, 21, 24
Rockwell International, 4, 5, 6, 79,
80, 87, 92
Rocky Flats Nuclear Weapons
Plant, **79-99**, 114, 117, 120,
123, 125, 126, 129, 130
Russell Sage Foundation, 3

Safety belts. *See* seat belts
Savannah River, Georgia, 79, 84
Seat belts, 7, 64, 68-69
Seveso, 103, 116, 130
Seveso Directive, 116
"Shadow government," 125
Shrivastava, Paul, 126
Silent spring, 1
Simulators, 117
Social responsibility, 2, 128

Standard-setting, 16-17, 19-20, 50-
51, 55, 64-65, 77-78, 105,
106, 125
Starr, Chauncey, 1
Swedish Motor Vehicle Inspection
Company, 77
Systems approach, 64

Taxonomy of hazards, 11, 13
3-M Corporation, 126
Three Mile Island, 103, 113, 123,
126, 130
Threshold limit values (TLVs), 19,
106 (table)
TLVs. *See* Threshold limit values
Tolerable risk, 24, 52, 53
Toxic Substances Control Act
(TSCA), 30, 31, 130-131, 131
Trade unions, 127-128
TSCA. *See* Toxic Substances Con-
trol Act (TSCA)
Tylenol, 123, 128

Union Carbide of India Limited
(UCIL), 105, 107
Union Carbide Corporation (UCC),
4, 5, 6, **101-117**, 120, 121,
123, 124

Voelz, George, 91-93
Volvo Car Corporation, 4, 5, 6,
57-78, 120, 124, 125, 128

WASH-1400. *See Reactor safety
study*
Weinberg, Alvin, 98

Zimmerman, Rae, 124

About the Book and Authors

This book provides one of the first systematic accounts of how corporations manage risk to workers and consumers. Careful analysis and interviewing in different corporations elicit a portrait of the myriad hazards that currently confront industry, the corporate programs and resources that have emerged since 1970 to respond to this challenge, and factors that have contributed to successes and failures.

In-depth studies of the Volvo Car Corporation, Rocky Flats nuclear weapons plant, Union Carbide's Bhopal facility, and large chemical and pharmaceutical corporations provide a state-of-the-art assessment of the advances and problems inherent today in industrial safety management.

Roger E. Kasperson, Jeanne X. Kasperson, and Christoph Hohenemser are senior researchers at Clark University's Center for Technology, Environment, and Development (CENTED). Roger E. Kasperson is Director of CENTED and its Hazard Assessment Group. Jeanne X. Kasperson is Research Librarian at CENTED and Senior Research Associate in Brown University's World Hunger Program. Hohenemser, a professor of physics, directs the Environment, Technology, and Society Program at Clark University. Robert W. Kates, formerly with CENTED, is Director of the Alan Shawn Feinstein World Hunger Program at Brown University. Ola Svenson, of the Department of Psychology at Lund University in Sweden, is a leading researcher in the field of risk perception and decision analysis.